PLANETARY CLIMATES

PLANETARY CLIMATES

Andrew P. Ingersoll

PRINCETON UNIVERSITY PRESS   *Princeton & Oxford*

Copyright © 2013 by Princeton University Press
Published by Princeton University Press, 41 William Street, Princeton, New
Jersey 08540
In the United Kingdom: Princeton University Press, 6 Oxford Street, Wood-
stock, Oxfordshire OX20 1TW

press.princeton.edu

ISBN 978-0-691-14504-4
ISBN (pbk.) 978-0-691-14505-1

Library of Congress Control Number: 2013939167

British Library Cataloging-in-Publication Data is available

This book has been composed in Minion Pro and Aviner LT Std

This book is printed on recycled paper ♻

Printed in the United States of America

10 9 8 7 6 5 4 3 2 1

Contents

PLANETARY CLIMATES

1 INTRODUCTION: THE DIVERSITY OF PLANETARY CLIMATES

CLIMATE IS THE AVERAGE WEATHER—LONG-TERM properties of the atmosphere like temperature, wind, cloudiness, and precipitation, and properties of the surface like snow, glaciers, rivers, and oceans. Earth has a wide range of climates, but the range among the planets is much greater. Studying the climates of other planets helps us understand the basic physical processes in a larger context. One learns which factors are important in setting the climate and how they interact.

Earth is the only planet with water in all three phases— solid, liquid, and gas. Mars has plenty of water, but it's almost all locked up in the polar caps as ice. There's a small amount of water vapor in its atmosphere but no standing bodies of liquid water. Venus has a small amount of water vapor in its atmosphere, but the Venus surface is hot enough to melt lead and is too hot for solid or liquid water. Thus by human standards, Venus is too hot and Mars is too cold. The classic "habitable zone," where Earth resides and life evolved, lies in between.

Things get strange in the outer solar system. Titan, a moon of Saturn, has rivers and lakes, but they're made of methane, which we know as natural gas. The giant planets have no solid or liquid surfaces, so you would need a balloon or an airplane to visit them. The climates there range from terribly cold at the tops of the clouds to scorching hot in the gaseous interiors, with warm, wet, rainy layers in between. Some of the moons in the outer solar system have oceans of liquid water beneath their icy crusts. The solar system's habitable zone could be an archipelago that includes these icy moons, but the crusts could be tens of kilometers thick. Their subsurface oceans are beyond the scope of this book.

The diversity of planetary climates is huge, but the basic ingredients are the same—the five elements H, O, C, N, and S. A fundamental difference is the relative abundance of hydrogen and oxygen. In the inner solar system—Earth, Mars, and Venus—the elements are combined into compounds like oxygen (O_2), carbon dioxide (CO_2), nitrogen (N_2), sulfur dioxide (SO_2), and water (H_2O). In the outer solar system—Jupiter, Saturn, Uranus, and Neptune—the elements are combined into compounds like methane (CH_4), ammonia (NH_3), hydrogen sulfide (H_2S), and water. Saturn's moon Titan has an atmosphere of nitrogen and methane, and Jupiter's moon Io has an atmosphere of sulfur dioxide. The composition of a planetary atmosphere has a profound effect on its climate, yet many of the processes that control the composition are poorly understood.

The underlying physical processes are the same as well. Temperature is a crucial variable, and it is largely but not entirely controlled by distance to the Sun. The temperature of the planet adjusts to maintain thermal equilibrium—to keep the amount of outgoing infrared radiation equal to the amount of absorbed sunlight.

Clouds and ice reflect sunlight, leading to cooler temperatures, but clouds also block outgoing infrared radiation, leading to warmer temperatures down below. Many gases like water vapor, carbon dioxide, methane, ammonia, and sulfur dioxide do the same. They are called greenhouse gases, although an actual greenhouse traps the warm air inside by blocking the wind outside. An atmosphere has nothing outside, just space, so the greenhouse gases trap heat by blocking the infrared radiation to space. Venus has clouds of sulfuric acid and a massive carbon dioxide atmosphere that together reflect 75% of the incident sunlight. Yet enough sunlight reaches the surface, and enough of the outgoing radiation is blocked, to make the surface of Venus hotter than any other surface in the solar system. Gases like nitrogen (N_2) and oxygen (O_2) do not block infrared radiation and are not significant contributors to the greenhouse effect.

The wind speeds on other planets defy intuition. At high altitudes on Venus, the winds blow two or three times faster than the jet streams of Earth, which blow at hurricane force although they usually don't touch the ground. In fact, Earth has the slowest winds of any planet in the solar system. Paradoxically, wind speed seems to increase with distance from the Sun. Jupiter has jet streams that blow three times faster than those on Earth, and Neptune has jet streams that blow ten times faster.

The weather is otherworldly. At least it is unlike what we are used to on Earth. Mars has two kinds of clouds—water and carbon dioxide. And Jupiter has three kinds—water, ammonia, and a compound of ammonia and hydrogen sulfide. Mars has dust storms that occasionally enshroud the planet. Jupiter and Saturn have no oceans and no solid surfaces, but they have lightning storms and rain clouds that dwarf the largest thunderstorms on Earth. Saturn stores its energy for decades and then erupts into a giant thunderstorm that sends out a tail that wraps around the planet.

Many of these processes are not well understood. Our Earth-based experience has proved inadequate to prepare us for the climates we have discovered on other planets. The planets have surprised us, and scientists often emerge from a planetary encounter with more questions than answers. But surprises tell us something new, and new questions lead to new approaches and greater understanding. If we knew what we would find every time a spacecraft visited a planet, then we wouldn't be learning anything. In the chapters that follow, we will see how much we know and don't know about climate, using the planets to provide a broader context than what we experience on Earth.

We will visit the planets in order of distance from the Sun, starting with Venus and ending with planets around other stars. Most planets get one or two chapters. Usually the first chapter is more descriptive—what the planet is like and how it got that way. The second chapter is more mechanistic—describing the physical processes that control the present climate of that planet. The chapters are augmented by sections called boxes, which contain equations and constitute a brief textbook-type introduction to climate science.

Chapter 2 is about the greenhouse effect and climate evolution, for which Venus is the prime example. Chapter 3 is about basic physical processes like convection, radiation, Hadley cells, and the accompanying winds, with Venus as the laboratory. Mars illustrates the "faint young Sun paradox," in which evidence of ancient rivers (chapter 4) contradicts results from astronomy that the Sun's output in the first billion years of the solar system was 70% of its current value. Mars also allows us to talk about the fundamental physical processes of condensation and evaporation (chapter 5), since exchanges of water vapor and CO_2 between the atmosphere and polar ice determine the climate of Mars. Titan allows us to study a hydrologic cycle in which the working fluid is not water (sections 6.1–6.3). Titan is an evolving atmosphere, close to the lower size limit of objects that can retain a sizeable atmosphere over geologic time (section 6.4). Below this limit, the atmospheres are tenuous and transient (section 6.5).

Jupiter is almost a cooled-down piece of the Sun, but the departures from solar composition tell a crucial story about how the solar system formed (chapter 7). The giant planets are laboratories for studying the effect of planetary rotation on climate (chapter 8), including the high-speed jet streams and storms that last for centuries. Chapter 9 is about Saturn, a close relative of Jupiter, although the differences are substantial and hard to understand. Uranus spins on its side, which allows us to compare sunlight and rotation for their effects on weather patterns (section 10.1). Neptune has the strongest winds of any planet (section 10.2), and we speculate about why this might be. The field of exoplanets—planets around other stars (section 10.3) is full of new discoveries, and we only give a brief introduction to this rapidly expanding field.

This book was written for a variety of readers. One is an undergraduate science major or a nonspecialist scientist who knows little about planets or climate. This reader will learn a lot about the planets and something about the fundamental physical processes that control climate. We go fairly deep into the physical processes, but the emphasis is on intuitive understanding. We touch on convection, radiation, atmospheric escape, evaporation, condensation, atmospheric chemistry, and the dynamics of rotating fluids. There are good textbooks and popular science books on planetary science[1,2,3,4] and there are multiauthored specialized books about individual planets.[5,6,7,8,9] There are also good textbooks on atmospheric science.[10,11,12,13] Therefore another potential reader is a student of atmospheric science who has learned the relevant equations and wants to step back and think about the fundamental processes in a broader planetary context. Finally, there are the climate specialists and planetary specialists who want to know about the mysteries and unsolved problems in planetary climate. Such readers might solve some of the many mysteries about planetary climates and thereby help us understand climate in general.

2 VENUS: ATMOSPHERIC EVOLUTION

2.1 EARTH'S SISTER PLANET GONE WRONG

UNTIL THE BEGINNING OF THE SPACE AGE, VENUS[5] WAS considered Earth's sister planet. In terms of size, mass, and distance from the Sun, it is the most Earth-like planet, and people assumed it had an Earth-like climate—a humid atmosphere, liquid water, and warm temperatures beneath its clouds, which were supposed to be made of condensed water. This benign picture came apart in the 1960s when radio telescopes[14] peering through the clouds measured brightness temperatures close to 700 K. Also in the 1960s, the angular distribution of reflected sunlight—the existence of a rainbow in the clouds—revealed that they were made of sulfuric acid droplets.[15] The Soviet Venera probes showed that the atmosphere was a massive reservoir of carbon dioxide, exceeding the reservoir of limestone rocks on Earth. The U.S. Pioneer Venus radar images showed a moderately cratered volcanic landscape with no trace of plate tectonics.

We now know that Venus mostly has an Earth-like inventory of volatiles—the basic ingredients of atmospheres and oceans—but with one glaring exception, and that is water. Earth's ocean is three hundred times as massive as its atmosphere. Water is more abundant than all the other volatiles combined, including carbon dioxide, nitrogen, and oxygen. In contrast, Venus has only a small amount of water, and it is all in the atmosphere. Relative to the mass of the planet, the amount of water on Venus is $\sim 4 \times 10^{-6}$ times the amount on Earth. This raises some fundamental questions: Was Venus born dry, and if so, how? Or, was Venus born with an Earth-like inventory of water that it somehow lost? These questions are the theme of this chapter. In section 2.1 we review the volatile inventories for the terrestrial planets. In section 2.2 we discuss the possibility that Venus lost a large amount of water, perhaps an ocean's worth, over geologic time. The evidence is found in the isotopes of hydrogen—the anomalously high deuterium to hydrogen ratio, which could come about when the lighter hydrogen escaped the planet's gravity at a higher rate than the heavier deuterium. Finally, in section 2.3 we discuss how Venus might have lost its ocean while Earth held on. The effect of extra sunlight on Venus is amplified by feedback—the warmer it gets, the more water vapor enters the atmosphere, which makes it even warmer because water vapor is a potent greenhouse gas. Somewhere between Earth and Venus, the theory goes, an Earth-like planet cannot exist. Inside this critical zone, the oceans boil away and are eventually lost to space.

Venus comes closer to the Earth than any other planet—its orbit around the Sun is 72% the size of Earth's orbit. The planet itself is 95% the size of Earth, and its surface gravity is 90% of Earth's. The bulk densities are nearly the same. The density of a terrestrial planet tells us about the proportions of rock and metal inside. The metal, mostly iron, is denser and resides in the core. The rocks are oxidized metals, mostly silicon, magnesium, and iron combined with oxygen, and have densities one-third that of the metallic core. From the bulk densities, one infers that the mass of rock relative to the mass of metal is about 2.2 for Earth and about 3.0 for Venus. Thus Venus and Earth are rocky terrestrial planets with similar amounts of metals locked away in their metallic cores.

Given these similarities in size, distance from the Sun, and bulk composition, one might expect the two planets would have similar climates and similar inventories of the lighter elements H, O, C, N, and S. These elements are the main ingredients of atmospheres, oceans, and ice sheets, which are the basic elements of climate. In some ways this expectation is correct, but in other ways it fails miserably. Venus has a small amount of water, all in vapor form, a massive atmosphere of carbon dioxide, and a surface temperature of ~730 K (457 °C , 855 °F). Clouds of sulfuric acid enshroud the planet (fig. 2.1). The pressure at the ground is 92 bars, or 92 times the Earth's sea-level pressure. Since pressure (force per unit area) is the weight of overlying atmosphere (gravity g times mass per unit area), the mass per unit area of the atmosphere of Venus is about 100 times that of Earth when one takes the lower gravity of Venus into account. Table 2.1 gives the volatile inventories of Venus, Earth, and Mars.[16,17]

Figure 2.1. Venus whole disk in the ultraviolet. The clouds in the image are made of sulfuric acid droplets and are ~65 km above the planet's surface.
(Source: http://nssdc.gsfc.nasa.gov/photo_gallery/photogallery -venus.html#clouds)

Earth actually has a lot of carbon dioxide, but it's not in the atmosphere. Instead it resides mostly in limestone—calcium and magnesium sediments with chemical formulas like $CaCO_3$ and $CaMg(CO_3)_2$. These carbonate compounds have accumulated over the life of the Earth when igneous rocks (rocks formed from a

Table 2.1.
Mass of volatiles per mass of planet. *Atm* is the amount in the atmosphere only. *Total* includes the amount in the crust, mantle, polar ice, and oceans. Except for the numbers in parentheses, the table is adapted from tables 4.2 and 4.4 of de Pater and Lissauer (2010).[1] CO_2 *total* for Earth is from carbonates in the crust and mantle given in table IV of Donohue and Pollack (1983).[16] The mantle value is highly uncertain. H_2O total for Earth is for the oceans only. H_2O *total* for Mars is from the volume of the polar caps given in table 10 of Smith et al. (2001).[38] CO_2 *total* for Mars is the CO_2 in the atmosphere plus the CO_2 ice in the polar caps reported by Phillips et al. (2011).[34]

	Venus	Earth	Mars
H_2O *atm*	1.2×10^{-9}	$\leq 1.6 \times 10^{-8}$	$\leq 1.6 \times 10^{-12}$
H_2O *total*	1.2×10^{-9}	2.3×10^{-4}	$(3.5–5) \times 10^{-6}$
CO_2 *atm*	9.5×10^{-5}	5.0×10^{-10}	3.7×10^{-8}
CO_2 *total*	9.5×10^{-5}	$(5–15) \times 10^{-5}$	$(6–7) \times 10^{-8}$
N_2 *atm*	22×10^{-7}	6.7×10^{-7}	6.7×10^{-10}
O_2 *atm*	7×10^{-10}	2.1×10^{-7}	3.7×10^{-11}
^{40}Ar *atm*	6×10^{-9}	11×10^{-9}	0.57×10^{-9}

melt) containing calcium and magnesium have weathered—ground down and dissolved—and then combined with carbon dioxide in the oceans where they precipitate out. Calcium is found in many kinds of igneous rock, but a simple example is $CaSiO_3$. Dissolved in water, CO_2 is a weak acid, which helps to dissolve the rock. The net result is that the $CaSiO_3$ combines with CO_2 to make SiO_2 (silica) and $CaCO_3$ (calcium carbonate).

The silica precipitates out as quartz and as hydrated silica (clay). The calcium carbonate precipitates out as limestone. As CO_2 is released from inside the Earth in volcanoes and fumaroles (volcanic vents), and as silicate rocks are weathered, the limestone accumulates. Biological processes aid the precipitation. Some plankton and other species use calcium carbonate in their shells, and much of the limestone deposits on Earth are derived from the shells of marine organisms. Other plankton species use silica in their shells. When one adds up the limestone deposits on Earth, the mass of CO_2 sequestered is 60 to 180 times the total mass of Earth's atmosphere. The larger number includes estimates of carbonates that were subducted into the mantle, and it is highly uncertain. In any case the mass of CO_2 in carbonate rocks on Earth is comparable to—within a factor of two—the mass of CO_2 in the atmosphere of Venus. Formation of carbonates depends on liquid water, so the fact that the CO_2 on Venus is in the atmosphere and the CO_2 on Earth is in carbonate rocks is probably tied to the question of why only Earth has oceans. That question hinges on the greenhouse effect and climate, as we shall see.

Before discussing water, let us discuss two other gases whose abundances are comparable on Earth and Venus. The first is nitrogen (N_2), which makes up 3.5% of Venus's 92-bar atmosphere and 78% of Earth's 1-bar atmosphere. These percentages refer to the numbers of molecules, not their masses. The ratio of the atmospheric masses of nitrogen on Venus and Earth is 3.0, which means they are comparable. The other gas is argon, which, on Earth at least, is mostly ^{40}Ar. The number 40 refers to the mass of the nucleus, which is controlled by the number of protons and neutrons. ^{40}Ar is produced from the decay of radioactive potassium, ^{40}K. The amount of ^{40}Ar in a terrestrial planet atmosphere is a measure of the amount of radioactive potassium in the crust and the degree of outgassing—the extent to which the gaseous argon is released from the rock and conveyed to the surface. Relative to the planet's mass, the ratio of the mass of ^{40}Ar in the atmosphere of Venus to that of the Earth is ~0.5, which means the amount of outgassing on the two planets is comparable as well.

The message from CO_2, N_2, and ^{40}Ar is that Venus and Earth have comparable inventories of at least some of the lighter elements and compounds, including those that are important for climate. These are the volatile compounds—those capable of existing as a gas or liquid at ordinary planetary temperatures. We have been using the word "comparable" to mean within a factor of 2 or 3. When it comes to water, however, Venus and Earth are dramatically different. For the sake of this discussion, we only consider the oceans and atmospheres, leaving off possible water in the crust and mantle. The depth of Earth's oceans varies, but if one turned all the oceans into vapor it would press down on the surface with a pressure of 260 bars. In other words, the mass of Earth's oceans is 260 times the mass of the atmosphere. Venus has no oceans. Water is present in the Venus atmosphere, mostly as vapor but also as a component of the sulfuric acid clouds, which are a compound of SO_3 and water. The total atmospheric water is 30 ppm (parts per million). The mass of Venus water relative to the mass of the planet is $\sim 5 \times 10^{-6}$ times the comparable number for

Earth. This is an astonishingly small ratio for two sister planets.

2.2 LOSS OF WATER AND ESCAPE OF HYDROGEN

So why are the planets so different in this important respect? Did they start out with comparable amounts of water, until Venus lost all but a tiny fraction of its initial inventory? Or was Venus born much drier than Earth, even though in many other respects the two planets are similar? A clue, or perhaps just an interesting fact, is that the water on Venus is different from the water on Earth. Although water is always two hydrogen atoms and one oxygen atom, the distributions of the isotopes of hydrogen and oxygen may differ. Oxygen has three stable isotopes, ^{16}O, ^{17}O, and ^{18}O, and hydrogen has two, H and D, where D stands for deuterium. The nucleus of ordinary hydrogen H is a single proton; that of deuterium D is a proton and a neutron, and since protons and neutrons have about the same mass, an atom of D is twice as massive as an atom of H. Their chemistry is the same, but their masses are different. This means that a process whose rate depends on mass may fractionate the hydrogen—alter the D/H ratio from its initial value. In most parts of the solar system the D/H ratio—the number of D atoms relative to the number of H atoms—is 10^{-4} within a factor of 2 in either direction.[18] This number applies to Earth, the giant planets, comets, and probably to the protosolar nebula out of which the Sun and planets

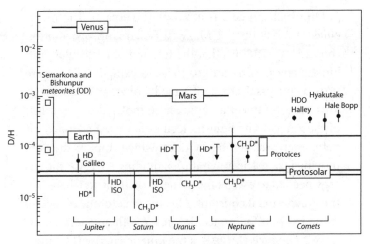

Figure 2.2. Deuterium to hydrogen (D/H) ratio in solar system. Asterisks are inferred from Earth-based observations. The high D/H of Venus suggests that the planet has lost a large amount of water during its lifetime.

(Adapted from fig. 4 of Bockelee-Morvan et al., 1998 and supplemented with material from Donohue et al., 1997)[18,19]

formed. Mars has a D/H ratio that is 7 times the Earth's ratio, indicating that some mass-dependent process has driven off the H at a faster rate than it has driven off D. However Venus has a D/H ratio that is at least 150 times the Earth's value[19] (fig. 2.2). This number applies to the water vapor in the lower atmosphere and to the hydrogen in the sulfuric acid clouds. Values up to 300 apply in the upper atmosphere, where the atmosphere is merging into space.[20] The implication is that some process has altered the D/H ratio on Venus, since the rest of the solar system is so nearly uniform in this respect.

One such process is atmospheric escape. The molecules of a gas obey a Maxwell-Boltzmann distribution (box 2.1), which says that the fraction of molecules with kinetic energy E is proportional to $\exp(-E/k_B T)$, where k_B is a universal constant called Boltzmann's constant. This probability is the same for all molecules, regardless of their mass. The light molecules have just as much kinetic energy as the heavy ones, which implies they are moving faster. The gravitational binding energy—the energy needed to escape the planet—is GMm/R, where G is the gravitational constant, M is the mass of the planet, m is the mass of the molecule, and R is the planet's radius (box 2.2). Since GMm/R is the kinetic energy $(1/2)mv_{esc}^2$ needed to escape, the escape velocity v_{esc} is $(2GM/R)^{1/2}$. If a molecule reaches escape velocity at the top of the atmosphere, where collisions with the molecules above are unlikely, then it will escape. The probability of this happening is greater for the lighter molecules, since their gravitational binding energy is less. Equivalently, one can say that the probability of reaching the escape velocity is greater for the lighter molecules because they are moving faster.

Atmospheric escape can alter the D/H ratio of what remains. An atom of H has a higher escape probability than an atom of D because the mass of H is one-half the mass of D. With H escaping faster, the D/H ratio of the remainder builds up. If one starts with water with hydrogen isotopic ratios like most other solar system bodies, that is, D/H $\sim 10^{-4}$, the ratio will slowly increase over time. The more water is lost, the larger the ratio

Box 2.1.

Distribution of Velocities in a Gas

The molecules of a gas follow a Maxwell-Boltzmann distribution of velocities. The probability of a molecule having its x-component of velocity in the range from v_x to $v_x + dv_x$ is

$$f(v_x)\,dv_x = \left(\frac{m}{2\pi k_B T}\right)^{1/2} \exp\left(\frac{mv_x^2}{2k_B T}\right)dv_x$$

Here m is the mass of the molecule, T is the absolute temperature, and k_B is Boltzmann's constant, 1.38×10^{-23} J K^{-1}. Similar distributions hold for the other two components of velocity. The probabilities of the three components are independent, so the joint probability is proportional to their product. Since exponents add when one takes the product of exponentials, the joint probability is proportional to $\exp[-\frac{1}{2}\,m\,(v_x^2 + v_y^2 + v_z^2)/(k_B T)] = \exp[-E/(k_B T)]$, where E is the kinetic energy of the molecule. Each distribution is normalized such that the integral over the entire range of v_x from $-\infty$ to $+\infty$ is 1.0. The mean speed, obtained by multiplying $f(v_x)$ by v_x and integrating from $-\infty$ to $+\infty$, is zero. The average of the square of the speed, obtained by multiplying by v_x^2 and integrating from $-\infty$ to $+\infty$, is $k_B T/m$, so the typical speed associated with each component of velocity is $(k_B T/m)^{1/2}$. The average kinetic energy, which is $(1/2)$ m times the average square of the speed, is $k_B T/2$ for each component, so the average kinetic energy of the molecule is $(3/2)$ $k_B T$. The average kinetic energy is independent of the mass of the molecule. For water vapor at 273 K, the value of $(k_B T/m)^{1/2}$ is 355 m s^{-1}. The velocity of the molecules is comparable to the speed of sound, which means their kinetic energy on a microscopic scale is large. This is part of the internal energy of the gas, which is what we sense when the gas is warm. If the mean velocity of the molecules is zero, the gas will have no kinetic

---

(*Box 2.1 continued*)

energy on a macroscopic scale. The other part of the internal energy is associated with the molecule's rotational and vibrational degrees of freedom and is also independent of its mass. For instance, a mole (6.02×10^{23} molecules) of hydrogen has approximately the same amount of internal energy as a mole of nitrogen or oxygen.

---

becomes. Working backward from today's D/H ratio to infer how much water has been lost is uncertain because one doesn't know the fractionation factor—how much advantage the H had relative to D in its rate of escape. If the temperature of the upper atmosphere were low over the age of the planet, the H would have had a large advantage, and a high D/H ratio would build up quickly. On the other hand, if the temperature were high, the H would have a smaller advantage, and it would take more water lost to build up a high D/H. The minimum amount of water lost from Venus is 150 times the current amount in the atmosphere, since that corresponds to no loss of D and only loss of H. That scenario is highly unlikely, so it is probable that much more water was lost, perhaps even an ocean's worth.

The above process is called thermal escape or equivalently Jeans escape. There are other escape processes, like the solar wind stripping atoms from the upper atmosphere and ultraviolet photons causing molecules to fly apart at speeds greater than the escape speed, and they all have the potential of increasing the D/H ratio.

Box 2.2.
Escape of Planetary Atmospheres

The escape velocity v_{esc} is the velocity that an object needs to escape the gravitational field of a planet. Equivalently, the kinetic energy $E_{esc} = (1/2)mv_{esc}^2$ is equal to the work done against gravity, i.e., the integral of the gravity force with respect to distance from the surface of the planet at radius R to a point infinitely far away:

$$E_{esc} = \int_R^\infty (GMm/r^2)\, dr = \frac{GMm}{R} = \frac{1}{2}mv_{esc}^2$$

Here G is the gravitational constant, M is the mass of the planet, m is the mass of the molecule, and v_{esc} is the escape velocity. If the molecule's kinetic energy is greater than E_{esc}, it will escape if it does not collide with other molecules. The above equation says that $v_{esc} = (2GM/R)^{1/2}$. Notice that v_{esc} is independent of the mass of the escaping particle. The escape velocity from the surface of Earth is 11.2 km s^{-1}. The escape velocity is less if the escaping particle has kinetic energy due to the planet's rotation or if the particle starts at some altitude above the planet's surface. For the terrestrial planets, only the molecules in the high-energy tail of the velocity distribution have enough energy to escape. The probability (box 2.1) of a molecule having kinetic energy E_{esc} is $\exp[-E_{esc}/(k_B T)] = \exp(-\lambda_{esc})$, where $\lambda_{esc} = GMm/(Rk_B T)$ is called the escape parameter. When λ_{esc} is large, e.g., when the mass of the molecule is large, the probability of escape is small. Also, when the mass of the planet is large and the radius is small, or when the temperature is small, then the probability of escape is small as well. This type of escape—from the high-energy tail of the velocity distribution, is called thermal escape or Jeans escape after Sir James Jeans, who first worked out the theory.

The large value of this ratio is telling us about the history of water on Venus, since water is where the hydrogen resides. Thermal escape does not remove intact water molecules—they are too heavy—but it does remove the atoms of D and H that are knocked off the water molecules when they absorb ultraviolet light, a process called photodissociation. The oxygen remains behind and either accumulates in the atmosphere or else combines with crustal materials to make oxides. For instance, iron in igneous rocks typically exists in ferrous form, FeO, but it can be oxidized to the ferric form, Fe_2O_3, thereby removing one oxygen atom for every two atoms of iron. Or unoxidized volcanic gases like CO, CH_4, H_2S, and H_2 can combine with oxygen to make CO_2, SO_2, and H_2O. Removal of oxygen from the atmosphere allows the hydrogen to escape, because otherwise the oxygen would build up and recombine with the hydrogen to make water. Whether Venus lost an entire ocean's worth of water by this process is unknown. What we do know is that the D/H ratio on Venus is two orders of magnitude higher than on almost every other object in the solar system.

2.3 THE RUNAWAY GREENHOUSE

Why should Venus have lost all its water when Earth did not? It could be that an Earth-like planet, with a liquid water ocean, can only exist outside a certain distance from the Sun. Venus could be inside this limit and Earth could be outside. The argument hinges on the extreme sensitivity of the climate to the amount of sunlight. The sensitivity arises because water vapor is a potent greenhouse gas and its abundance is controlled by cycles of evaporation and precipitation. On Earth, these cycles maintain the relative humidity—the amount of water vapor the air does hold compared to the amount it could hold—at an average value of about 50%. As the air warms, it holds more water vapor. Venus, being closer to the Sun, is naturally warmer. But that leads to more water vapor in its atmosphere, and that makes it even warmer, since water vapor is a potent greenhouse gas. The intensity of sunlight is about twice as great at the orbit of Venus compared to the orbit of Earth, and this may be enough to have pushed Venus over the edge into its current hellish state.

A personal anecdote is appropriate here. In the first class I ever taught, I composed a homework problem with a twist. I asked the students to compute the temperature at the surface of a planet that absorbed a certain fraction of the incident sunlight and whose atmosphere blocked a certain fraction of the outgoing infrared radiation. The blocking was due to greenhouse gases, and the point was to illustrate the greenhouse effect. With enough simplifying assumptions, it is a fairly easy problem, but I made the amount of greenhouse gases depend on temperature. The cycles of evaporation and precipitation would ensure that the atmosphere was 50% saturated with water vapor, so higher temperatures meant more water vapor, and more water vapor meant higher temperatures because water vapor is a potent greenhouse gas.

The homework problem didn't work. The Earth couldn't exist in its present state, according to the assumptions given. I apologized to the class and dialed down the sensitivity to the surface temperature as much as I could. Then the Earth could exist but Venus couldn't, at least not with an atmosphere that was close to saturation. The only way for Venus to exist was to make it so hot that the air was no longer saturated, but that meant boiling away the oceans. I wrote up the results and sent the paper off to a reputable scientific journal, where it was rejected as too speculative. So I added a lot of features to make it more realistic, and I sent it off to another reputable journal. This time it was rejected because the added features didn't measure up to the standards of terrestrial meteorology. I couldn't meet those standards because I was trying to guess what Venus was like four billion years ago, so I stripped the model down to its essence and explained that it was the concept I was after and not the quantitative details. The third journal, which was also quite reputable, accepted the paper, and it was published as "The Runaway Greenhouse: A History of Water on Venus."[21]

In engineering, feedback is a return of a portion of the output of a device back to its input. If the device is the climate of a planet and the output is temperature, the inputs are the absorbed sunlight and the emitted infrared radiation. If either of these depends on temperature, then you have feedback. Negative feedback tends to stabilize the system. Outgoing long-wave radiation is an example: The warmer it gets, the more infrared radiation is emitted, and this cools the planet back to its equilibrium state. Positive feedback tends to destabilize the system, and two examples come to mind. The first is ice-albedo feedback, where the ability of the planet to absorb sunlight depends on the temperature. Albedo is the fraction of the incident sunlight that gets reflected back to space. The colder it gets, the more ice accumulates on the ground, which causes more sunlight to be reflected. This cools the Earth further and leads to large climate excursions and possibly even to tipping points—regime changes—where the system changes to a new stable state. A cold, ice-covered Earth—called snowball Earth by the geologists—would stay cold by reflecting the incoming sunlight. There is geological evidence that such a regime occurred several times in early Earth history.

The positive feedback that is most relevant for Venus is due to water vapor. If one took an Earth-like planet with a global ocean and moved it closer to the Sun, the water-vapor content of the atmosphere would rise and amplify the warming effect of the extra sunlight. At some orbital position, probably between the orbits of Earth and Venus, the climate system would run away. The planet would have a massive water vapor atmosphere with a surface pressure of 300 bars, most of it due to the weight of water vapor, and the rest—the original nitrogen and oxygen—accounting for an additional 1 bar. For a watery planet moving in from Earth's orbit, this transition is a tipping point, a regime change. For a watery planet at the orbit of Venus, it is the only possible state of the system.

What we have just described is called a runaway greenhouse,[21] to distinguish it from the modest terrestrial

greenhouse that keeps our planet warm and comfortable. Climate models[22] have reproduced the runaway
greenhouse state for Venus, with surface temperatures
as high as 1400 K. If Venus started wet but immediately
went into a runaway greenhouse state, the water would
no longer be protected from photodissociation by ultraviolet light from the Sun. Photodissociation occurs in
the stratosphere and above, where the ultraviolet photons first encounter the atmospheric gases. On Earth
these high-altitude layers are predominantly nitrogen
and oxygen. Water vapor is present in abundances of 1
or 2 parts per million. In particular, the oxygen forms a
radiation shield that protects Earth's water from ultraviolet light and prevents the chain of events that leads to
hydrogen escape and oxidation of crustal materials. On
a watery planet at the orbit of Venus, water vapor would
be the major constituent at all altitudes.[21] Instead of 1 or
2 photons striking a water molecule out of one million
incident photons, almost all of the photons would do
so. This could lead to destruction of water and escape of
hydrogen over geologic time, and could account for the
absence of water on Venus today.

The near-zero abundance of water on Venus today
is on firm ground. The high D/H ratio is also on firm
ground. The existence of water vapor feedback is a
theoretical concept, but it is also on fairly firm ground.
However the whole history of water on Venus remains
uncertain. It is still possible that Venus was born dry,
although this contradicts the sister planet evidence, especially the similarities in the abundances of CO_2, N_2,

and ^{40}Ar. It is also possible that the high D/H reflects relatively modern processes. Small reservoirs of hydrogen, like the present Venus atmosphere, are easier to alter than large reservoirs like the Earth's oceans. This doesn't necessarily imply that Venus was born dry, but it does complicate the interpretation of the high D/H ratio.

Could a runaway greenhouse happen on Earth? The incident sunlight is almost two times greater at Venus, so a runaway greenhouse is more likely there. We could increase the odds for Earth by plugging up the infrared spectrum of our atmosphere with greenhouse gases. Their heat-trapping effect—the radiative forcing—is qualitatively the same as increasing the sunlight, but we have a long way to go. The radiative forcing due to human activities up to the present is in the 1% range.[10] A factor of two seems safely out of reach, but we don't know where the tipping point lies. It is probably between the orbits of Earth and Venus, since Earth has an ocean and Venus does not. We don't know exactly how Venus got to be such a hellish place, but it is a lesson for Earth nonetheless.

3 VENUS: ENERGY TRANSPORT AND WINDS

···

3.1 CONVECTION

To this point we have invoked the greenhouse effect in fairly general terms, but it is time to discuss the modern-day climate of Venus in greater detail. The details matter if we want to compare Venus to Earth to see how likely a runaway greenhouse is for our planet. The essence of the atmospheric greenhouse is that the gas is more transparent to sunlight than it is to infrared radiation. The sunlight that reaches the surface is absorbed and turned into heat, but the infrared radiation can't carry it up to the levels where it is radiated to space—the gas is too opaque. Instead, the surface heats up until convection kicks in. Convection involves warm air moving up and cold air moving down, leading to a net transfer of energy upward. Convection only kicks in where the lapse rate reaches a certain critical value, where lapse rate is defined as the rate of temperature decrease with altitude. Once the critical point is reached, however, convection is very efficient in transporting heat upward. The atmosphere reaches a kind of radiative-convective equilibrium, where sunlight deposits heat at the surface and convection carries the heat up to the levels where the gas becomes transparent to infrared radiation, at which point the heat is radiated to space. With its massive atmosphere, Venus radiates to space at altitudes around 65 km. Earth radiates to space at altitudes around 5 km. It's like piling more blankets on the bed. The thicker the pile, the warmer you are at the bottom.

Most planets radiate to space from altitudes where the pressure is several hundred mbar. This is the level where the typical infrared photon stands an even chance of passing through the layers above without absorption. It's the level you "see" when you peer into an atmosphere with infrared eyes. You see the warm glow of the gases at that level. If they are cold, you see very little glow; if they are warm, the glow is greater. The pressure you see depends on how transparent or opaque the gases are at infrared wavelengths. Certain gases, like water vapor and CO_2 are good absorbers of infrared light, which means you can't see as deep when these gases are present. These are the greenhouse gases—they block infrared light and raise the altitude (lower the pressure) at which the planet radiates to space. By raising the altitude, they are increasing the thickness of the pile of blankets, and this raises the surface temperature. Clouds also absorb infrared light. Venus has a uniform layer of high clouds and an atmosphere of CO_2, whereas Earth has a ~50% cloud cover and an atmosphere with small amounts of water vapor and CO_2. Therefore one sees deeper into Earth's atmosphere—to 500 mbar—than into Venus's atmosphere, where one sees only to ~100 mbar.

Box 3.1.
Gases in Hydrostatic Equilibrium

The equation of state of an ideal gas is often written $PV = Nk_B T$, where P is pressure, V is volume, N is the number of molecules, k_B is Boltzmann's constant (see box 2.1), and T is the absolute temperature. One can divide by volume to get $P = nk_B T$, where $n = N/V$ is the number of molecules per unit volume. One can multiply and divide by the mass m of the molecule and use the definition of density $\rho = mn$ to get

$$P = \rho k_B T/m = \rho R_g T$$

Here $R_g = k_B/m$ is the gas constant for that particular gas. If there is a mixture of gases, then $m = (\Sigma m_i n_i)/(\Sigma n_i)$ is the average of the masses weighted by the number of molecules of each type.

Hydrostatic equilibrium is when the pressure difference dP between the bottom and top of a layer supports its weight. Let z be the vertical coordinate. If dz is the thickness of the layer, the mass per unit area is ρdz and the weight per unit area is $\rho g dz$. The force per unit area is dP, which is equal to $-\rho g dz$ when the layer is in hydrostatic equilibrium. The minus sign signifies that the pressure increases as the height decreases. Thus the equation for hydrostatic equilibrium is

$$dP/dz = -\rho g$$

We can substitute for the density using the equation of state $\rho = P/(R_g T)$ to get

$$\frac{1}{P}\frac{dp}{dz} = -\frac{g}{R_g T} = -\frac{1}{H}$$

Here $H = R_g T/g = k_B T/(mg)$ is called the scale height. If temperature and gravity were constant, pressure and density would vary as $\exp(-z/H)$, and H would be the vertical distance over which the pressure decreases by a factor of $e = 2.718\ldots$ More generally, pressure varies as $\exp(-\int dz/H)$, but in all cases the scale height is a useful measure of atmospheric thickness. For the midtroposphere of Earth ($T = 255$ K, $g = 9.8$ m s^{-2}, and $m = 0.029$ kg per mole), one finds that $H = 7.5$ km.

How high is the 100-mbar level on Venus? How thick is the pile of blankets? The surface pressure on Venus is 92 bar—92 times the sea level pressure on Earth. If we define a scale height (box 3.1) as the vertical distance over which the pressure drops by a factor of $e = 2.718\ldots$, then there are $ln(92/0.1) = 6.8$ scale heights from the surface to the 100 mbar level. On Earth there are $ln(1/0.5) = 0.69$ scale heights from the surface to the 500 mbar level. Because the pressure range is so much greater for Venus than for Earth, the height where the planets radiate to space is much greater as well. The scale height is proportional to temperature, and ranges from 6 to 9 km on Earth and from 5 to 15 km on Venus. Venus radiates to space from altitudes near 65 km. Earth radiates to space from altitudes near 5 km.

Figure 3.1 shows the temperature of the Venus atmosphere from the ground to an altitude of 80 km.[23] It is a vertical profile measured at one location on the planet, but it is typical of the planet as a whole because the massive atmosphere, with its large heat-carrying capacity, tends to reduce the horizontal variations. We will treat this profile as an average for the planet, with the sunlight

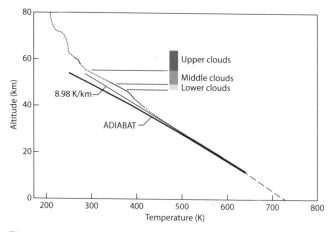

Figure 3.1. Venus temperature variation with altitude from the Pioneer Venus large probe. The profile is nearly the same as a global average. Contrasts from day to night and equator to pole are less than ±10 K. Altitudes where the temperature follows an adiabat are those where convection is occurring. (Adapted from fig. 11 of Seiff, 1983)[23]

averaged over night and day and over high and low latitudes. On Venus, 75% of the incident sunlight is reflected back to space, mostly by the sulfuric acid clouds and the dense CO_2 atmosphere. The remaining 25% is absorbed, both in the atmosphere and at the surface. To maintain thermal equilibrium, all of the absorbed energy must be transported upward to the level where the photons can escape to space.

Figure 3.1 shows the measured profiles and other curves labeled adiabats. An adiabat is the pressure versus temperature relation for a parcel that moves without exchanging heat with its surroundings (box 3.2).

Box 3.2.
Adiabatic Lapse Rate and Stability

The lapse rate is the decrease of temperature with height. The actual lapse rate compared to the adiabatic lapse rate tells whether the atmosphere is stable to convection or not. The adiabatic lapse rate follows from the first law of thermodynamics. For an ideal gas the change in internal energy is $C_v dT$. Therefore

$$C_v dT + P dV = dQ$$

Here C_v is the specific heat at constant volume and dQ is the added heat, which is zero for an adiabatic process. The volume per unit mass is V, which is $1/\rho$, so $dV = (R_g/P)dT - (R_g T/P^2)dP$, which follows from the equation of state. Therefore

$$(C_v + R_g)\, dT - (R_g T/P)\, dP = dQ$$

The first term is the heat added per unit mass when $dP = 0$. Thus $C_v + R_g$ is C_p, the specific heat at constant pressure. The second term is $-(1/\rho)dP$, which follows from the equation of state, and $(1/\rho)dP$ is $-g dz$, which follows from the equation for hydrostatic equilibrium. Thus for an adiabatic process, i.e., for $dQ = 0$, we have

$$C_p dT + g dz = 0, \quad \text{or} \quad \frac{dT}{dz} = -\frac{g}{C_p}$$

This equation gives the temperature change with respect to height when an air parcel moves adiabatically in a hydrostatic gas. The parcel cools as it goes up because it is expanding and doing work on its surroundings. If dT/dz of the surroundings is more positive than $-g/C_p$, then the parcel, rising adiabatically, will find the surroundings warmer than itself. The parcel will be more dense than the surroundings and it will fall back.

(*Box 3.2 continued*)

In the opposite case, when dT/dz of the surroundings is more negative than $-g/C_p$, then the surroundings will be colder and more dense than the parcel and it will continue to rise. In this case the atmosphere is unstable, since a small displacement tends to grow. The atmosphere is stable when $dT/dz + g/C_p$ is positive, and it is unstable when $dT/dz + g/C_p$ is negative. This derivation assumes there is no condensation, so g/C_p is called the dry adiabatic lapse rate.

The adiabats all have the same slope, such that $dT/dz = -g/C_p$, but they have different intercepts, that is, different temperatures at a given pressure level. Here g is the gravitational acceleration and C_p is the specific heat of the atmosphere. Thus there are warm adiabats and cold adiabats. The decrease of temperature with altitude is called the lapse rate, so g/C_p is called the adiabatic lapse rate. Convection is occurring where the lapse rate is equal to the adiabatic value. Convection cannot occur when the lapse rate is less than the adiabatic lapse rate, such that temperature falls off more gradually than the adiabat ($dT/dz > -g/C_p$). Then a parcel displaced rapidly upward will find itself colder than its surroundings. Its lower temperature means higher density than the surroundings, and the parcel will sink back to its original level. Thus an atmosphere with a sub-adiabatic lapse rate is stable, and convection will not occur.

On the other hand, if the atmospheric temperature falls off more steeply than the adiabat ($dT/dz < -g/C_p$), a parcel displaced upward will find itself warmer and less dense than the surroundings, and it will continue to rise. Since it is warmer, it will carry extra energy with it. Similarly, a sinking parcel will continue to sink and it will carry a deficit of energy with it. Thus an atmosphere with a super-adiabatic lapse rate, where $dT/dz < -g/C_p$, is unstable to convection. Convection implies an upward transfer of energy, since hot air is rising and cold air is sinking. The net effect of convection is to warm the upper levels and cool the lower levels. This makes dT/dz less negative—it reduces the unstable lapse rate and brings it closer to the adiabat, at which point the convection shuts itself off. In practice, convection is so efficient that the lapse rate never exceeds the adiabatic value by a significant amount. When one sees the temperature profile following an adiabat, one can infer that convection is occurring and energy is being transported upward. When the lapse rate is sub-adiabatic, one can infer that convection is absent. According to figure 3.1, convection is occurring below 25 km and within the clouds from 47 to 56 km.

3.2 RADIATION

In this one-dimensional view of the Venus atmosphere, the only other process that transports energy vertically is electromagnetic radiation (box 3.3).

Acting together, radiation and convection largely determine the temperature structure of the Venus atmosphere. In steady state, each layer of the atmosphere

Box 3.3.
Electromagnetic Radiation: Flux and Intensity

The radiant energy flux (irradiance), is the power per unit area falling on a surface. An example is the flux of energy in solar radiation. At 1 AU from the Sun the average flux is 1361 $W\,m^{-2}$, and it varies by $\pm 1\ W\,m^{-2}$ due to solar activity. This is the power you would receive if you held a square meter perpendicular to the sunlight. As you move around in the solar system, the flux varies as $1/r^2$, where r is distance to the Sun. One can regard this inverse-square dependence in two ways. One way is that the same amount of radiant power passes through each sphere of radius r surrounding the Sun. The power per unit area on the sphere is inversely proportional to the sphere's area, which is $4\pi r^2$. The other way is that the power that you receive is proportional to the area of the Sun in the sky. The Sun looks smaller as you move away from it. It's the angular area, measured in square radians, that counts. This area, called solid angle, is $\pi(R/r)^2$, where R is the Sun's radius. The inverse square dependence arises from the factor of r^2 in the denominator. Implicit in this argument is that there is some intrinsic property of the Sun that is independent of r. This intrinsic property is the intensity (also called radiance or brightness) of the solar disk, and it is the flux per unit solid angle in a narrow beam. The brightness of the solar disk is the same wherever you are, but the solid angle varies as $1/r^2$. An example is a blank wall. The wall looks just as bright from a large distance as it does from a small distance. The brightness is an intrinsic property of the wall. Solid angle is measured in square radians, or steradians, and is abbreviated sr. Just as the total angular size of a circle is 2π (the circumference divided by r), the total solid angle of a sphere is 4π (the sphere's surface area divided by r^2).

Flux is the integral of the intensity over all directions. For example, a horizontal surface radiates into the upward hemisphere, whose total solid angle is 2π. In spherical polar coordinates, the element of solid angle is $\sin\theta\, d\theta\, dl$ where θ is the angle from the polar axis and l is longitude. Each element of solid angle, times the intensity in that direction, multiplied by the cosine of the angle with respect to vertical, contributes to the vertical flux. For the special case of a surface whose intensity I is the same in all directions, the flux F is π times the intensity:

$$F = \int_0^{2\pi} \int_0^{\pi/2} I \cos\theta \sin\theta\, d\theta\, dl = \pi I$$

A snowfield is a good approximation. It has the same brightness whether you look at it obliquely or straight down. Another example is a small hole in a large furnace whose inner walls are all at the same temperature. You see the same red-hot glow (brightness) regardless of the angle, although the hole looks smaller (as $\cos\theta$) when you look at it obliquely. The red-hot glow is called cavity radiation, or blackbody radiation, and is described further in box 3.4.

adjusts its temperature until it is gaining as much energy as it is losing. The gain may be from absorption of sunlight, from absorption of infrared radiation emitted by other parts of the atmosphere, or from warm parcels rising up from below, replacing cold parcels that sink down. The loss may be from emission of infrared radiation or from exchange of warm parcels for cold ones.

A useful approximation for treating radiative transfer involves the concept of blackbody radiation (box 3.4). Blackbody radiation exists inside a closed cavity

Box 3.4.

Blackbody Radiation, the Planck Function

Inside an isothermal cavity, the radiation field comes into equilibrium with the walls. A hole in the walls does not disturb the light inside if the hole is small enough. Any light that goes in through the hole gets lost inside and doesn't find its way out, so the hole is described as black. The light coming out of the hole originates inside and is called blackbody radiation, or cavity radiation. The intensity inside is a universal function of wavelength and temperature called the Planck function, after Max Planck who derived the exact shape of the function by assuming that the energy of each optical mode could change only in quantized steps proportional to the frequency of the light. Frequency ν is c/λ, where λ is the wavelength and c is the velocity of the light. Planck's discovery was the beginning of quantum mechanics, and the constant of proportionality h, Planck's constant, is the fundamental constant of quantum mechanics. The Planck function depends on h, c, and Boltzmann's constant k_B, which is the fundamental constant of statistical mechanics. For blackbody radiation, the intensity (power per cross-sectional area per steradian) within a wavelength interval from λ to $\lambda + d\lambda$ is

$$B_\lambda(T)\,d\lambda = \frac{2(k_B T)^4}{h^3 c^2}\frac{w^{-5}dw}{e^{1/w}-1}, \quad w = \frac{\lambda k_B T}{hc} = 69.504\,\lambda T$$

Here λ is in meters, T is in Kelvins, and w is a dimensionless variable proportional to wavelength. See figure 3.2 for a graph of the function $w^{-5}/(e^{1/w}-1)$. The integral over w from 0 to ∞ is an integral over the entire spectrum of blackbody radiation. After multiplying by π the result is

$$F_{BB} = \pi\int_0^\infty B_\lambda(T)\,d\lambda = \frac{2\pi(k_B T)^4}{h^3 c^2}\int_0^\infty \frac{w^{-5}dw}{e^{1/w}-1} = \sigma T^4$$

The factor of π was inserted to convert intensity into flux assuming the intensity is independent of direction, as it is for blackbody radiation (box 3.3). Thus σT^4 is the total flux, integrated over all wavelengths, from a blackbody at temperature T. It is the flux coming out of the small hole in the cavity. The Stefan-Boltzmann constant $\sigma = 5.67 \times 10^{-8}$ W m^{-2} K^{-4} is a universal constant that depends only on h, k_B, and c. For a 273 K blackbody, the flux is 315 W m^{-2}.

Figure 3.2 shows the dimensionless Planck function $w^{-5}/(e^{1/w} - 1)$. The abscissa is the dimensionless wavelength w. The peak of the dimensionless Planck function is at $w = 0.2014$, so if $T = 5800$ K, then λ at the peak is 0.50 microns (0.5 $\times$ 10^{-6} m). These numbers provide a good fit to the Sun's spectrum, which has total energy about equal to a 5800 K blackbody and a peak at $\lambda = 0.5$ microns—in the green portion of the spectrum.

whose walls are all at the same temperature, where the radiation field has come into equilibrium with the walls. Blackbody radiation also exists inside a thick isothermal medium like a cloud or an absorbing gas. Remarkably, many real substances behave like blackbodies even if you take away half of the medium—the other half emits as if it were still surrounded by opaque cloud or gas. Then the outgoing heat flux (power per unit area) is σT^4, where T is its absolute temperature, and σ is a universal physical constant—the Stefan-Boltzmann constant. This universality applies to a perfectly absorbing (nonscattering) medium, but most planetary substances come close to this ideal at infrared wavelengths (box 3.5).

Box 3.5.
Blackbodies in the Solar System

It is remarkable that the Sun, which is pouring energy out into space, should have the same distribution of radiation as a closed cavity that is in thermal equilibrium with the walls. The Sun behaves this way because the ions and electrons in the Sun's atmosphere are interacting with each other so frequently that they remain in thermal equilibrium despite the exposure to empty space. The Sun is a good blackbody, and its 5800 K brightness temperature is roughly that of the ions and electrons in the Sun's atmosphere.

Blackbody radiation is relevant in planetary science because most planetary materials are good absorbers of infrared light. In other words, they behave like blackbodies at infrared wavelengths. Infrared is the relevant wavelength because that is where the Planck function peaks at planetary temperatures. For a temperature of 290 K, which is 1/20 of the Sun's effective temperature, the peak is at a wavelength of 20 times the peak of the solar spectrum, or 10 microns. This follows from the definition of w (box 3.4), which says that λ scales as $1/T$ for the same value of w. Light at 10 microns is in the middle of the thermal infrared, and that is where most planetary materials, including rocks, sand, ice, liquid water, clouds, and gases, behave as blackbodies. The same materials reflect visible light, so they are not such good absorbers at those wavelengths. Metals are not good absorbers in the infrared, so they do not radiate as blackbodies. On the other hand, metals are not important elements of planetary climate.

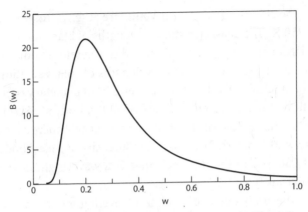

Figure 3.2. Planck function in dimensionless units. The Planck function $B_\lambda(T)$ gives the distribution of blackbody radiation with respect to wavelength. The graph shows the universal function $B(w) = w^{-5}/(e^{1/w} - 1)$, where w is proportional to wavelength and temperature. In dimensional units $B_\lambda(T)d\lambda$ is $B(w)dw$ times a factor proportional to T to the fourth power. See box 3.4 for a precise definition of variables.

The spectrum of blackbody radiation is given by the Planck function $B_\lambda(T)$, which is a function only of the wavelength λ and the temperature T (box 3.4). The dimensionless form of the Planck function is shown in figure 3.2. Blackbody radiation is thermal radiation, which occurs when the energy of molecular motions is turned into electromagnetic energy. Sunlight is thermal radiation, as is the infrared radiation emitted by the planets. The difference is that the particles in the Sun's atmosphere—atoms, ions, and electrons—are at an average temperature of 5800 K, and particles near the tops of the clouds on Earth and Venus are at temperatures of ~250 K. The characteristic wavelengths of blackbody radiation are inversely proportional to the temperature of the emitting body, so the wavelengths of solar radiation are much shorter than the wavelengths of planetary radiation. The maximum intensity of a blackbody at 5800 K is in the middle of the visible range at 0.5 microns, or 0.5×10^{-6} meters. The maximum intensity of a blackbody at 250 K is in the infrared range at a wavelength of 11.6 microns (box 3.4).

The probability of a photon traveling a certain distance decreases exponentially as the distance increases (box 3.6). The probability is usually expressed in terms of optical depth τ_λ, such that $\exp(-\tau_\lambda)$ is the transmission—the probability of the photon going the distance. Optical depth is an integral over distance. The integrand depends on the medium—its density ρ and the ability of a unit mass to absorb light. The latter is called the absorption coefficient k_λ, and it can be a strong function of the wavelength λ. One sees farther into the medium at wavelengths where the absorption coefficient is low. For a nonscattering medium, the intensity of radiation is a weighted average of $B_\lambda(T)$ from all the points along the ray path from $\tau_\lambda = 0$ to $\tau_\lambda \gg 1$ (box 3.7). The weighting is proportional to $\exp(-\tau_\lambda)$, the probability that the photons can reach the observer without being absorbed.

As mentioned earlier, the measured temperature profile on Venus follows a dry adiabat below 25 km and within the clouds from 47 to 56 km (fig. 3.1). These layers are places where the infrared radiation is unable to

Box 3.6.
Radiative Transfer

Consider a narrow beam of radiation propagating through a planetary atmosphere. The light encounters a thin slab of gas that is aligned perpendicular to the direction of propagation, and the slab has thickness ds. A fraction of the intensity will be absorbed, another fraction will be scattered, and the remaining fraction will be transmitted through the slab. In addition, the slab will emit thermal radiation, which will increase the intensity exiting the slab. Scattering is a change in the direction of the light without loss of energy. Emission is a conversion of internal energy of the medium into electromagnetic energy. The extinction—the fraction that is either absorbed or scattered—is called the optical thickness, which is written $d\tau_\lambda = k_\lambda \rho\, ds$, where ρ is the density of the gas and k_λ is an intrinsic property of the gas called the extinction coefficient. The subscript λ indicates that the extinction coefficient is a function of the wavelength of the light. The difference between the intensity emerging from the slab and the intensity incident on the other side is

$$dI_\lambda = I_\lambda^{out} - I_\lambda^{in} = -\, d\tau_\lambda I_\lambda + dJ_\lambda$$

The first term on the right is the extinction. The second term on the right has two parts. One is the light scattered into the beam from other directions, and the other is thermal emission. The scattered light depends on the incident radiation—the radiation environment of the slab—and the thermal emission does not. We derive the expression for dJ_λ by imagining the slab at equilibrium inside an isothermal cavity. Then $I_\lambda = B_\lambda$ everywhere, and $dI_\lambda = 0$, which means $dJ_\lambda = d\tau_\lambda B_\lambda$. For the special case of a gas that does not scatter radiation, dJ_λ is independent of the radiation environment, so it remains the same when we take the slab out of the isothermal cavity. Therefore

(Box 3.6 continued)

$$dI_\lambda = -d\tau_\lambda(I_\lambda - B_\lambda), \quad \text{or} \quad \frac{dI_\lambda}{d\tau_\lambda} = -I_\lambda + B_\lambda$$

This is the equation of transfer for a nonscattering (pure absorbing) medium. Extinction and absorption are the same for this case. The minus sign assumes that τ_λ increases in the direction of the light beam, although we will make the opposite assumption later. Radiative transfer in a purely absorbing medium is simply the slab removing intensity $d\tau_\lambda I_\lambda$ from the beam and replacing it by $d\tau_\lambda B_\lambda$ where B_λ is the Planck function at the temperature of the slab.

Box 3.7.
Solution of the Equation of Transfer

The equation of transfer is linear in the intensity and can be solved for $I_\lambda(\tau_\lambda)$ when $B_\lambda(\tau_\lambda)$ is given. Here $\tau_\lambda = \int k_\lambda \rho \, ds$ is a coordinate that is zero at the observer and increases backward along the ray path to a point where the optical depth is τ_λ^* and the intensity is $I_\lambda(\tau_\lambda^*)$. Thus $I_\lambda(\tau_\lambda^*)$ is the incident intensity entering the slab from the other side. The slab may be optically thin ($\tau_\lambda^* \ll 1$), optically thick ($\tau_\lambda^* \gg 1$), or anything in between. The Planck function depends on τ_λ because temperature is a function of position within the medium. The solution to the equation of transfer is

$$I_\lambda(0) = I_\lambda(\tau_\lambda^*)\exp(-\tau_\lambda^*) + \int_0^{\tau_\lambda^*} B(\tau_\lambda)\exp(-\tau_\lambda)\,d\tau_\lambda$$

The exponential in the first term is the transmission—the fraction of the incident light that reaches the observer. Since the absorption coefficient of the medium is wavelength dependent, the transmission is also wavelength dependent. The second term is the sum of the contributions $d\tau_\lambda B_\lambda$ from slabs of infinitesimal thickness, each contribution multiplied by the exponential transmission between its source and the observer. A simple case is when the whole slab is optically thick ($\tau_\lambda^* \gg 1$) and the medium is isothermal at temperature T. Then $B_\lambda(\tau_\lambda)$ is a constant, $B_\lambda(T)$, and can be moved outside the integral, which goes from 0 to infinity. The result is $I_\lambda(0) = B_\lambda(T)$, which means that the slab emits as a blackbody.

carry the vertical heat flux. The upward flux of infrared radiation is actually a net flux—the difference between upward-moving photons and downward-moving photons. At a given level, the upward and downward fluxes are weighted averages of the Planck function of the layers below and above, respectively. If the product of absorption coefficient and gas density is large, the weighted averages will extend only short distances into the medium. Then the upward and downward fluxes will each be equal to the blackbody flux σT^4 at the local temperature T (box 3.8). The net flux will be zero, and the radiation will not be able to carry the necessary heat flux up from below, where it was deposited by sunlight. Convection has to step in. Apparently the absorption coefficient is large within the clouds because the clouds are opaque to infrared radiation. The absorption coefficient times the density is large below 25 km because the gas density is high.

If there is a temperature gradient within the medium, the heat flux from the hot side is greater than that from

Box 3.8.
Optically Thick Atmosphere

Imagine we are immersed deep within a medium that is optically thick in all directions, either because it is physically thick or because the absorption coefficient is large—the medium is opaque at the wavelength of interest. The transmission factor (box 3.7) in the integral for $I_\lambda(0)$ says that we "see" into the medium to optical depths of order unity. This will be a small physical distance if the absorption coefficient is large. Over that small distance it may be possible to expand the Planck function $B_\lambda(\tau_\lambda)$ as a constant plus a gradient term:

$$B_\lambda(\tau_\lambda) = B_\lambda(0) + \tau_\lambda \frac{dB_\lambda(0)}{d\tau_\lambda} = B_\lambda(0) + \frac{\tau_\lambda}{k_\lambda \rho} \frac{dB_\lambda(0)}{ds}$$

The second step follows because $d\tau_\lambda = k_\lambda \rho ds$. Here s is distance back into the medium, opposite to the direction of propagation. $dB_\lambda(0)/ds$ is a constant independent of τ_λ. Substituting this expression into the solution of the equation of transfer (box 3.7) with $\tau_\lambda^* \gg 1$, one obtains

$$I_\lambda(0) = B_\lambda(0) \int_0^\infty \exp(-\tau_\lambda) d\tau_\lambda + \frac{1}{k_\lambda \rho} \frac{dB_\lambda(0)}{ds} \int_0^\infty \tau_\lambda \exp(-\tau_\lambda) d\tau_\lambda$$

$$= B_\lambda(0) + \frac{1}{k_\lambda \rho} \frac{dB_\lambda(0)}{ds}$$

If the ray is propagating upward at an angle θ with respect to the vertical, then $dB_\lambda(0)/ds$ is $-\cos\theta\, dB_\lambda(0)/dz$, since $ds = -dz/\cos\theta$ is opposite to the direction of propagation. If the ray is propagating downward at an angle θ with respect to the vertical, then $dB_\lambda(0)/ds$ is $\cos\theta\, dB_\lambda(0)/dz$. Thus

$$I_\lambda^{up}(0) = B_\lambda(0) - \frac{\cos\theta}{k_\lambda \rho} \frac{dB_\lambda(0)}{dz}, \quad I_\lambda^{down}(0) = B_\lambda(0) + \frac{\cos\theta}{k_\lambda \rho} \frac{dB_\lambda(0)}{dz}$$

Highly oblique rays ($\cos\theta \ll 1$) see less deep into the medium and therefore sense less of the gradient than rays propagating vertically ($\cos\theta = 1$). To get the upward and downward fluxes, one multiplies by $\cos\theta$ again and then integrates over all solid angles $\sin\theta\,d\theta\,dl$ in the upward and downward hemispheres (box 3.3). The flux spectral density—the flux per unit wavelength interval—is

$$F_\lambda^{up} = B_\lambda(0) - \frac{2}{3k_\lambda\rho}\frac{dB_\lambda(0)}{dz}, \quad F_\lambda^{down} = B_\lambda(0) + \frac{2}{3k_\lambda\rho}\frac{dB_\lambda(0)}{dz}$$

The net upward flux is the difference $F_\lambda^{up} - F_\lambda^{down} = F_\lambda^{net}$. The leading terms cancel because to first order, one sees the same temperature in all directions. One gets a net flux only when the photons can sense a gradient of temperature. Integrating over wavelength one gets the total net flux:

$$F_\lambda^{net} = -\int_0^\infty \frac{4\pi}{3k_\lambda\rho}\frac{dB_\lambda(0)}{dz}\,d\lambda = -\left|\frac{4\pi}{3\rho}\int_0^\infty \frac{1}{k_\lambda}\frac{dB_\lambda}{dT}\,d\lambda\right|\frac{dT}{dz}$$

In an optically thick medium radiative heat transfer looks like thermal conduction. The quantity in square brackets is called the radiative conductivity. It is an average of $1/k_\lambda$ weighted by dB_λ/dT, a function whose wavelength dependence resembles that of the Planck function (Box 3.4 and Fig. 3.2). The resulting average is called the Rosseland mean. The strongest contributions to the radiative conductivity are from wavelengths where the gas is most transparent, i.e., where $1/k_\lambda$ is large.

the cold side, so there is a net flux from hot to cold (box 3.8). The net flux is greater if the medium is more transparent, because then the photons travel farther in the medium and sense greater temperature contrasts. If the medium is not too transparent, the radiative heat flux is proportional to the temperature gradient. Then radiative heat transfer behaves like thermal conduction, and the thermal conductivity is inversely proportional to the absorption coefficient times the density. This is because the photons travel farther when the absorption is low, which means they can carry a greater heat flux for the same temperature gradient (box 3.8). As long as infrared radiation can carry the required upward heat flux with a temperature gradient that is less than the adiabatic lapse rate, convection will not occur. Apparently the atmosphere from 25 to 47 km is sufficiently transparent to do this.

3.3 RADIATIVE-CONVECTIVE EQUILIBRIUM

Not only each layer, but also the planet as a whole must be in thermal equilibrium. This means that the levels at which the photons escape to space must have the right temperature to balance the absorbed sunlight. We will refer to this altitude range as the emitting level, although really it includes all levels for which the optical depth τ_λ measured from outside the atmosphere, is of order unity (box 3.9). The altitude of the emitting level depends on the composition of the atmosphere—specifically on the absorption coefficient, which depends on the abundance of greenhouse gases and on the opacity of the clouds. On Venus the pressure at the emitting level is ~0.1 bars. On Earth the pressure at the emitting level is ~0.5 bars.[10] Earth's atmosphere is more transparent to infrared radiation, so photons from higher pressures can escape to space; the altitude at the emitting level is only 5 km.

Box 3.9.
Emission to Space

Emission to space is the way planets get rid of the heat that they absorb from the Sun. The altitude of the emitting levels depends on the radiative properties of the atmosphere like the absorption coefficient and optical depth. The temperature of the emitting levels adjusts to keep the planet in thermal equilibrium and depends on the absorbed sunlight. Scientists have measured the emission to space for all the planets, first to find out how they achieve thermal equilibrium and second to find out what their atmospheres are made of, as revealed in their infrared spectra.

Consider a spacecraft measuring the outgoing longwave (infrared) radiation that is propagating straight up from below. We define the optical depth at the spacecraft to be zero, with optical depth increasing downward into the atmosphere. For this example, let the optical depth of the ground be large, i.e., $\tau_\lambda^* \gg 1$. The altitude of the spacecraft is z_{sc}. Then the solution to the equation of transfer (box 3.7) can be written

$$I_\lambda(0) = \int_0^\infty B_\lambda(\tau_\lambda)\exp(-\tau_\lambda)\,d\tau_\lambda = \int_{-\infty}^{z_{sc}} B_\lambda(z)\left[\exp(-\tau_\lambda)\,k_\lambda\rho\right]dz$$

The second step follows because $d\tau_\lambda = -k_\lambda\rho dz$. The quantity in square brackets on the right is called the weighting function $W_\lambda(z)$. Its integral with respect to z is unity. The weighting function gives the contribution of the Planck function within each altitude range $(z, z + dz)$ to the intensity at the top of the atmosphere. $W_\lambda(z)$ goes to zero at the top of the atmosphere because the gas density ρ goes to zero. $W_\lambda(z)$ goes to zero deep in the atmosphere because the transmission to space $\exp(-\tau_\lambda)$ goes to zero. The weighting function has a single maximum in between, near the level where $\tau_\lambda \approx 1$. Values of $W_\lambda(z)$ greater

---

(Box 3.9 continued)

than half the peak value extend at least one scale height above and below (see box 3.1 for a definition of scale height). Since τ_λ is the integral of ρk_λ with respect to depth, the peak of the weighting function ($\tau_\lambda \approx 1$) is deeper in the atmosphere at wavelengths where the absorption coefficient is weaker.

On Venus the emitting level is at a much higher altitude, 65 km, because there is so much more atmosphere underneath.

The blackbody flux from the emitting levels must be able to balance the sunlight absorbed by the planet. The emitted flux includes all the altitudes from which the photons are able to escape to space, that is, where $\exp(-\tau_\lambda)$ is not too small. The average emitted flux includes the altitudes sampled at different wavelengths. The global average includes all places on the planet as well. If we express this global average flux as a blackbody flux at a single equilibrium temperature, the expression is $T_e = [(1 - A)F_o/(4r^2\sigma)]^{1/4}$. Here A is the planet's albedo for visible light, the fraction of incident sunlight that is reflected to space. Thus $(1 - A)$ is the fraction of incident light that is absorbed. F_o is the energy flux of sunlight (1361 W m^{-2}) at Earth orbit, and r is the ratio of the planet's orbital radius to Earth's orbital radius, which is called the Astronomical Unit (AU). The factor of 4 is the ratio of the surface area of a sphere to the cross-sectional area of a sphere (box 3.10). For Venus, $r = 0.72$, $A = 0.75$, and $T_e = 231$ K. For Earth, $r = 1.0$, $A = 0.29$, and $T_e = 255$ K.

Box 3.10.
Equilibrium Temperature

The weighting function (box 3.9) gives the range of altitudes from which emission to space occurs. The peak of the weighting function occurs where the optical depth measured from the top of the atmosphere is unity. We call this the emitting level; its altitude is controlled by the density of the gas and its absorption coefficient. If the planet is in equilibrium, the temperature at the emitting level is controlled by the amount of sunlight absorbed and any internal heat that the planet is losing. Internal heat is important for the giant planets, but we will neglect it here. Let F_o be the energy flux of sunlight (1361 W m^{-2}) at Earth orbit and r be the ratio of the planet's orbital radius (distance to the Sun) to Earth's orbital radius, which is called the Astronomical Unit. The total solar power absorbed by a planet is the product of three factors—the power per unit area in the incident sunlight F_o/r^2, the cross-sectional area of the planet πR^2, and the fraction of sunlight that is absorbed $(1 - A)$. Here A is the planet's albedo for visible light and R is the planet's radius at the emitting level. Thus the power absorbed is $(1 - A)\pi R^2 F_o/r^2$. We express the power emitted as $4\pi R^2\sigma T_e^4$, where T_e is the equilibrium temperature—the temperature the planet would have in equilibrium if it radiated as a blackbody uniformly over its surface area of $4\pi R^2$. Thus

$$T_e = \left[\frac{(1 - A)F_0}{4\sigma}\right]^{1/4}$$

The factor of 4 is the ratio of the surface area of the sphere to the cross-sectional area.

The emitting level is really a range of altitudes ~2 scale heights thick (box 3.9) where the optical depth τ_λ is of order unity. The question then arises, what is the temperature at the

Venus has a lower value of T_e than Earth despite being closer to the Sun because the Venus albedo is so much higher.

For Venus, the massive atmosphere distributes the heat fairly evenly, so the actual temperature near the 65 km level at different points on the planet is close to T_e, or 231 K. If the atmosphere were convective throughout, from 65 km down to the surface, the surface temperature would be 810 K. This calculation assumes that temperature increases with depth at the adiabatic lapse rate $g/C_p = 8.9$ K km^{-1}. Here $g = 8.9$ m s^{-2} and $C_p = 1000$ J K^{-1} kg^{-1}, which are appropriate for Venus's CO_2 atmosphere. The calculation overestimates the surface temperature because the lapse rate is less than the dry adiabatic lapse rate at certain altitudes—below the clouds from 25 to

47 km and in the upper cloud from 56 to 65 km altitude. Thus the average lapse is 7.7 K km, which gives the observed surface temperature of 730 K.

There are several ways one could increase the surface temperature of Venus. One way is to plug up the spectral windows where the atmosphere is partially transparent to infrared radiation. By making the atmosphere opaque at all levels, one could eliminate the subadiabatic layers from 25 to 47 km and from 56 to 65 km. Then the atmosphere would be convective, the lapse rate would be adiabatic, and the surface temperature would be 810 K, as discussed earlier. This could be done by adding greenhouse gases to the atmosphere, especially gases that absorb at different wavelengths than those covered by CO_2. Another way is to raise the level of emission to space by adding clouds and greenhouse gases above 65 km altitude. This would raise all the temperatures below if the lapse rate remained the same. A third way is to decrease the albedo, which would raise the equilibrium temperature T_e, which would become the new temperature at the 65 km level. A fourth way is to allow more sunlight to penetrate to the ground, forcing the atmosphere to carry a larger heat flux up to the emitting level. This might exceed the ability of radiation to carry the heat in the sub-adiabatic layers from 25 to 47 km and 55 to 65 km, forcing them to become adiabatic. This raises the surface temperature to 810 K as before.

At the highest levels, where $\tau_\lambda \ll 1$, most of the photons are escaping. These layers are sandwiched between outer space, which emits essentially nothing, and the layers below, which emit at a rate (power per unit area) of σT_e^4, where T_e is the equilibrium temperature of the planet, about 231 K for Venus. Unless they are absorbing sunlight, the layers where $\tau_\lambda \ll 1$ assume the average T^4 of the two sides of the sandwich. Thus $T^4 = (0 + T_e^4)/2$, or $T = T_e/2^{1/4}$, which is about 194 K for Venus. This is the simplest way to estimate the stratospheric temperature of a planet (box 3.10). For Earth the same calculation gives 214 K for the stratosphere. Above the stratosphere, where the gas density is extremely low, even a small amount of sunlight can raise the temperature. Earth has a layer where ozone is abundant, and that gas causes a bulge in the atmospheric temperature. On Venus, which has only a small amount of ozone, the temperature remains low to very high altitudes.

Emission to space allows one to estimate the composition of the atmosphere by remote sensing (box 3.11). Since the altitude of the emitting level depends on the absorption coefficient, it depends on the wavelength of the light. At wavelengths where the absorption coefficient is small, one sees deeper into the atmosphere, and at wavelengths where the absorption coefficient is large, one sees less deep. Usually the strong absorptions are concentrated in narrow spectral lines. If temperature increases with depth, the spectral lines will appear dark relative to the radiation outside the lines, which comes from warmer, deeper levels. Most gases have numerous spectral lines, whose exact wavelengths constitute a unique fingerprint of the absorbing gas. This is the basis of solar spectroscopy and of thermal infrared spectroscopy of planets.

Box 3.11.
Spectroscopy

One sees deeper into the atmosphere at wavelengths where the absorption coefficient is weak (box 3.9). If the gases in the atmosphere have spectral lines where the absorption is especially strong, one sees higher in the atmosphere. If there is a temperature gradient, e.g., if temperature increases with depth, then one will measure lower intensity in the spectral lines—they will look darker. This is how the Fraunhofer lines in the solar spectrum originate, and it is how scientists estimate the composition of the Sun's atmosphere. Planetary atmospheres are the same, except the thermal radiation comes out in the infrared where the molecular vibrations and rotations of gases leave their spectral signatures. The noble gases and diatomic gases like N_2, O_2, and H_2 are harder to see in the infrared because they have no pure vibrational and rotational absorptions.

The spectrum of thermal radiation is only half the story, however. The other half is the spectrum of short-wave radiation that was emitted by the Sun and reflected off the clouds or the surface of a planet. The gases in the atmosphere leave their spectral signatures as the sunlight penetrates into the atmosphere and then out to our detector. The darkness of the spectral lines gives the amount of gases along the light path. If one knows the path, one can estimate the abundance. Problems arise if the photons bounce around inside the cloud, which lengthens the path and leads to overestimating the abundance unless corrections are made. Nevertheless, the spectrum of reflected sunlight is just as powerful a tool for measuring the composition as the spectrum of thermal emission.

The atmospheric greenhouse effect depends on sunlight reaching the ground and infrared not being able to transfer it out. Convection has to take over, which steepens the lapse rate and warms the surface. Venus is special because the atmosphere is so massive. Emission to space occurs at an extremely high altitude, which forces the surface temperature to its extremely high value. Essentially, one is piling more blankets on the bed.

3.4 HADLEY CELLS

So far we have treated the atmosphere as if it had only one dimension. We discussed vertical heat transfer in what is called a radiative-convective equilibrium model. When developed fully, such models give a fairly accurate estimate for the role of greenhouse gases and clouds in regulating climate. However, these models neglect the horizontal variations in the amount of sunlight—the difference between day and night, equator and poles, winter and summer. Venus is like Earth in that the Sun is nearly overhead at the equator and low on the horizon at the poles. Actually the effect is more pronounced at Venus because the spin axis is almost exactly perpendicular to the orbit plane, whereas Earth's spin axis is tilted at 23°. The tilt angle is called the obliquity, and the obliquity causes the seasons—the north pole is in sunlight for half the year and the south pole is in sunlight for the other half. Venus has a 177° obliquity—the spin axis is within 3° of being opposite to the orbital motion around the Sun. Because this angle is small, Venus has almost no seasons, and the poles get very little sunlight. Averaged over a year, the incident sunlight is more concentrated at low latitudes on Venus than it is on Earth. The fact that the planet spins backward probably does not affect the climate in a significant way. The massive atmosphere and the length of the day (117 Earth days) are more important factors.

Because the planet is spinning, the absorbed sunlight is spread out in longitude. Like the colors of a spinning top, the pattern is symmetric about the axis of rotation. The massive atmosphere is able to store the absorbed sunlight during the day and radiate it at night without a large drop in temperature. As a result of this heat storage, the greatest thermal contrasts are between the different latitudes and not between the day and night sides. As on Earth, the equator is generally hotter than the poles. However, the massive atmosphere is able to transport heat effectively from equator to poles, which tends to reduce thermal contrasts in that direction as well. A fluid transports heat away from a given region by carrying more energy when it leaves than when it returns. The fluid parcels are like circulating dump trucks. The dump trucks don't accumulate at the dump, but whatever they are carrying does.

Perhaps the simplest type of circulation that transports heat poleward is rising motion at the equator and sinking motion at higher latitudes. The higher temperatures at the equator cause the air to expand, which raises the pressure at the higher altitudes. The higher pressure pushes the air away from the equator, so a pattern develops with rising motion at the equator and sink-ing motion to the north and south of the equator. The spreading fluid carries more energy than the returning fluid, because there is more gravitational potential en-ergy at higher altitude (box 3.12). The larger amount of gravitational potential energy is partly offset by less ther-mal energy, since the air is colder at high altitude, but the sum of the two increases with altitude in a stably strati-fied atmosphere. As we have seen, the Venus atmosphere is stably stratified (convection is absent) in the altitude range from 25 to 47 km (fig. 3.1).

In a constant gravitational field the gravitational po-tential energy per unit mass is gz. The thermal energy per unit mass is $C_p T$. Their sum, called dry static energy (box 3.12), is $C_p T + gz$. A fluid parcel moving adiabati-cally—not exchanging heat with its surroundings—has a constant value of dry static energy even though T and z are changing. If $dT/dz > -g/C_p$, the dry static energy in-creases with altitude. As we have seen, this is the criterion that the air is stably stratified—that convection is absent. Thus the simple overturning circulation, with heated air rising at the equator and sinking at higher latitudes, is transporting energy poleward. The circulation is called the Hadley circulation, after George Hadley, an amateur meteorologist of the eighteenth century. Poleward mo-tion is observed out to 45° latitude in both the northern and southern hemispheres of Venus, but the observa-tions refer to the tops of the clouds—at around 65 km altitude[24] (fig. 3.3). The vertical extent of the poleward

Box 3.12.

Energy Transport by Fluid Motions

At low latitudes most planets absorb more energy than they radiate, and at high latitudes they radiate more energy than they absorb. This is because the Sun at noon is nearly overhead at low latitudes and is low on the horizon at high latitudes. These imbalances require energy transport from low to high latitudes. Consider an imaginary great wall that circles the planet at a constant intermediate latitude, say 30°. This wall extends from the ground into space, so the atmosphere must transport energy through the wall, not above or below it. On Earth, the oceans also transport energy. Fluid parcels passing through the wall carry energy with them, and they also do work on the fluid in front of them. If the parcels return from high latitudes with less energy than they left with, and if they do less work on the fluid in front as they return, then there is a net energy transport toward high latitudes.

The mass flux through the wall (mass/area/time) is ρv, where ρ is density and v is the velocity perpendicular to the wall. To see this, consider a piece of the wall with area dA. In time dt, the fluid moves a distance vdt, so a volume $dA\,vdt$ moves through dA in time dt. Multiplying by the density gives the mass moving through the area dA in time dt, so the mass per unit area per unit time is ρv. The energy flux through the wall is the mass flux times the energy per unit mass. For an ideal gas the internal energy per unit mass is $C_v T$ and the gravitational potential energy per unit mass is gz. We are treating C_v and g as constants. Therefore the power per unit area carried with the fluid is $(C_v T + gz)\rho v$. We combine this with the rate at which work per unit area is being done, which is force per unit area times velocity or Pv, where P is pressure. For an ideal gas,

(Box 3.12 continued)

this can be written $\rho R_g T v$, so the total rate of energy transport (power) through the wall is

$$\int [(C_v + R_g)T + gz]\rho v \, dA = \int (C_p T + gz)\rho v \, dA$$

The integral is taken over the entire surface area of the wall, and dA is an element of area. We have used $C_v + R_g = C_p$, which holds for ideal gases (box 3.2).

Energy is transported poleward even though the total rate of mass transport $\int \rho v dA$ is zero. Parcels move toward the pole at some locations on the wall, and they move away from the pole at other locations, but on average the mass flow in the two directions cancel out. Mass is not accumulating at any point on the planet. To transport energy, the poleward-moving parcels must have higher values of $C_p T + gz$ than the equatorward-moving parcels. This can happen if the fluid moving toward the pole is warmer than the returning flow or if it takes place at a higher altitude, or both. The quantity $C_p T + gz$ is called dry static energy. It increases with altitude in a stably stratified atmosphere (box 3.2), that is, when the lapse rate is less than the dry adiabatic value. Since atmospheres are mostly stable, the poleward flow must take place at high altitude and the return flow must take place at low altitude if energy is to be transported poleward. This is the circulation pattern of Earth's Hadley cell and the probable pattern of overturning circulations on the other planets.

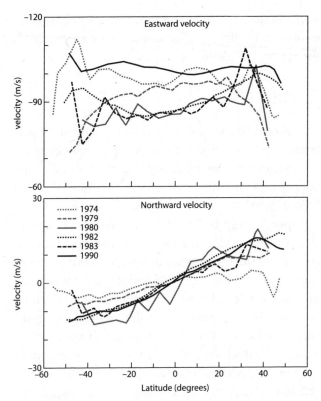

Figure 3.3. Venus winds at cloud-top level. The eastward winds are negative, meaning that they blow to the west. The northward winds blow toward the pole in each hemisphere.

(Adapted from fig. 3 of Gierasch et al., 1997)[24]

flow and the nature of the return flow are unknown. On Earth, the Hadley circulation goes out only to ±30°, so the Hadley circulation on Venus is wider.

3.5 EDDIES

The Hadley circulation is a zonally symmetric circulation. It is an averge with respect to longitude, so the overturning is the same around the latitude circle. Superposed on this symmetric circulation are circulations involving waves and eddies, and they too can transport energy. If the zonally symmetric circulation is steady—not varying in time—the eddy circulation is the part of the flow that varies in longitude. An example is northward currents alternating with southward currents around a circle of constant latitude. If the fluid moving northward is warmer than the fluid moving southward, there will be a net transfer of energy to the north, assuming the fluid motions are all at the same altitude (no difference in gravitational energy). More generally, a correlation between dry static energy and northward velocity represents a northward transfer of energy, which we call the eddy heat transport (box 3.13). The dump truck analogy is still useful. The fluid parcels are dumping their energy at the northern part of their circuit and returning to the south to pick up more energy. On Earth, this eddy heat transfer is dominant beyond ± 30°, which is the latitude limit of the Hadley circulation. On Venus, the handoff from symmetric circulation to the eddy circulation is uncertain. It has not been possible to measure

Box 3.13.

Energy Transport by Eddies

The Hadley cell is an average with respect to longitude. By definition, it is symmetric about the axis of rotation—independent of longitude. The symmetric circulation coexists with the eddies, which are fluctuations relative to the mean. Earth's rotation inhibits symmetric circulations at high and midlatitudes, so an eddy circulation takes over. This circulation varies in longitude, poleward flow alternating in longitude with equatorward flow. The former has higher values of dry static energy than the latter, as required for poleward energy transport (box 3.12), but this is accomplished mainly by having warm air moving poleward and cold air moving equatorward, without much difference in altitude. The word "eddy" means any departure from a steady symmetric circulation. Poleward eddy heat transport occurs when temperature and poleward velocity are correlated as they vary in longitude. The extratropical cyclones and anticyclones that constitute much of the weather at midlatitudes are the primary elements of the eddy heat transport on Earth.

Here the word "correlated" means that the two variables are in phase with each other. There is a theorem that involves the product of two fluctuating quantities—call them u and v. Let

$$u = \bar{u} + u', \quad v = \bar{v} + v', \quad \text{then} \quad \overline{uv} = \bar{u}\bar{v} + \overline{u'v'}$$

The overbar is the mean and the prime is the fluctuation from the mean. If the mean is taken with respect to longitude, then the fluctuation is the residual—the leftover part that varies with longitude. The fluctuating part has zero mean, by definition. Since u and v each involve two terms, the product uv involves four terms, but two of them, $\bar{u}v'$ and $u'\bar{v}$, are zero in the mean. This leaves the two terms on the right, the first representing

(*Box 3.13 continued*)

the product of the two means and the second representing the product of the two fluctuations. When the averages are with respect to longitude, the first is the symmetric part and the second is the eddy part. Even though the fluctuations have zero means, their product will be nonzero if the fluctuations are correlated.

the temperature and the northward component of velocity to adequately resolve the eddy heat flux. The best velocity measurements are obtained by visual tracking of clouds at 65 km altitude. A handful of parachuted probes and floating balloons have been tracked from Earth at lower altitudes, using the Doppler shift in the probes' radio signals to measure velocity.[25] The small numbers of these devices and the large separation in time (years or decades) makes for poor statistics, and it is hard to define the eddy circulation from these data. A fleet of small balloons, each measuring temperature and pressure and equipped with a radio transmitter, is probably required to understand the Venus climate.

Although we have not fully determined how the heat is transported poleward, we have good evidence that poleward transport is taking place (fig. 3.4). The evidence comes from satellite data that give the fluxes of absorbed sunlight and emitted infrared at the top of the atmosphere for individual latitude bands. The incident sunlight is known to high accuracy. The solar flux is 1361 $W\,m^{-2}$ at Earth orbit, and the value varies

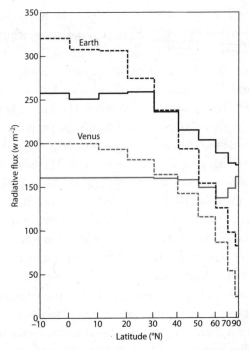

Figure 3.4. Absorbed sunlight (dashed) and emitted infrared radiation (solid). Distance along the x-axis is proportional to the sine of the latitude and is therefore proportional to area on the sphere measured from the equator. The emitted curve is flatter and the equator-to-pole temperature difference is smaller for Venus than for Earth.

(Adapted from fig. 11 of Taylor et al., 1983)[26]

throughout the solar system as $1/r^2$, where r is distance to the Sun. Satellite data give the albedo A at visible wavelengths, and $1 - A$ is the fraction that is absorbed. The dominant effect is the falloff of the absorbed sunlight with latitude, which is mainly due to the lower angle of the Sun above the horizon at higher latitudes. Satellite data also give the emitted infrared radiation, which falls off less steeply with latitude than the absorbed sunlight.

Figure 3.4 shows observations of the absorbed sunlight and the emitted infrared radiation for the northern hemispheres of Earth and Venus.[26] The units are power per unit area $(W\,m^{-2})$. The abscissa is proportional to the sine of the latitude, which is proportional to the area between that latitude and the equator. The advantage of this coordinate is that area under each curve is the total power (W) absorbed or emitted in the northern hemisphere. For both planets, the total absorbed power and the total emitted power are the same. Any small imbalances are lost in the noise of the measurement. On the other hand, the imbalances within each latitude band are large and easily measured. On both planets the equatorial latitudes are gaining net energy through the top of the atmosphere, and the high latitudes are losing net energy. Since each latitude has to remain in thermal equilibrium, there must be a transport of energy from low latitudes to high latitudes.

The most interesting fact about figure 3.4 is that the Venus infrared radiation is more nearly independent of latitude than that of the Earth. Since infrared emission is related to temperature, this means that the temperature of the Venus atmosphere at the emission level is more nearly independent of latitude than that of the Earth. The Sun is preferentially heating the equator and trying to create a temperature difference, but the atmospheric heat transport is trying to reduce the difference. On

Venus, the atmosphere is winning, and on Earth, it's a tie. This is partly due to the enormous heat-carrying capacity of the Venus atmosphere, which is more successful in eliminating temperature differences than Earth's atmosphere. A subtle question is why the Earth's oceans, which have more heat-carrying capacity than the Venus atmosphere, don't make a greater contribution to the poleward heat transport on Earth. Part of the answer is that sunlight does not penetrate more than a few tens of meters into the oceans. The oceans, therefore, are heated from above, which makes them stably stratified and resistant to overturning. Less than 10% of the ocean water is warm. Below depths of a few hundred meters, the oceans all over the world are close to the freezing point. In contrast, Venus has enough sunlight absorbed at the bottom to keep the atmosphere stirred by convection, at least up to the 25–47 km layer where convection stops. This stirring enables the atmosphere at all levels to take part in the global circulation.

The other part of the answer to why the Venus atmosphere is winning the contest is the slow rotation. Rotation gives rise to the Coriolis force, which acts on a moving body when we measure its motion in a rotating coordinate system. In a game of catch on a merry-go-round, the ball would appear to follow a trajectory curving to the right if the merry-go-round were rotating counterclockwise. The Coriolis force is proportional to the angular velocity of the planet and the speed of the ball, or the speed of the wind in a planetary atmosphere. It acts to the right of the velocity vector in the northern hemisphere. On a rapidly rotating planet like Earth, an eastward-moving jet stream develops at mid-latitudes, and its Coriolis force pushes back on the pressure force that is tending to drive the upper branch of the Hadley cell. An eastward-moving jet stream develops in the southern hemisphere as well, although there the Coriolis force is to the left of the wind, and the upper branch of the Hadley cell is to the south. This pushback leads to a balance of forces, with nothing left to drive the Hadley cell. On a slowly rotating planet like Venus, the Coriolis force is weaker, the balance is less complete, and the Hadley cell is stronger and wider. This leads to more heat transported poleward and a smaller equator-to-pole temperature difference, as seen in figure 3.4.

Because of its slow rotation, Venus is like the Earth's tropics, where the Coriolis force—actually its horizontal component—is weak. At the equator, the Coriolis force becomes a vertical force and is overwhelmed by gravity—its effect on the winds is negligible. In the Earth's tropics, the horizontal pressure gradients are small because they have almost nothing to push against—just a weak Coriolis force. Since pressure gradients are related to temperature gradients through the thermal expansion of the gas, the horizontal temperature gradients also are small. This is another way to look at the Venus atmosphere: It resembles the Earth's tropics, where temperature varies from place to place only by a small amount.[12]

The climate at the surface of Venus is hot, dry, and uniform from place to place. Temperature varies with altitude, but at a given altitude the temperature is nearly constant over the surface of the planet. The engineers in charge of the Soviet Venera spacecraft described the noontime brightness as being like Moscow on a cloudy winter day. The day lasts for 117 Earth days, but the Sun is never visible in the sky. Enough light filters down through the clouds to take photos with your camera, provided you somehow protected it from the searing heat. Winds at the surface have not been measured, but they are likely to be small. This uniformity is like that at the bottom of the oceans, where the temperature everywhere is just 1 or 2 °C. In both cases, the enormous mass of the overlying fluid ensures that temperatures are constant and that the bottom layers have settled into a low energy state. Nevertheless, radar images of the surface reveal occasional dunes and streaks downstream of topographic features. Tides driven either by the Sun's gravity or by the Sun's daily heating could drive winds in the lower atmosphere. Tides could be important in maintaining the unusual backward rotation of the planet. A long-lived (one Venus day) instrumented lander at the surface could address these issues, but such a mission is technically challenging because of the searing heat.

3.6 SUPERROTATION

The weather gets more interesting at the tops of the clouds—at 65 km altitude and above. There the winds are primarily zonal and blow to the west at speeds up to 100 m s^{-1} (figs. 3.3 and 3.5).[24] The 100 m s^{-1} speed is 2.5 times faster than the Earth's jet streams and is 40% of

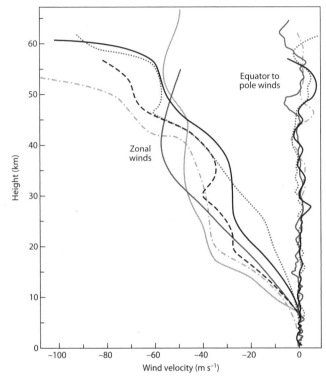

Figure 3.5. Vertical wind profile obtained by measuring the Doppler shift of radio signals from the Pioneer Venus and Venera probes. The Pioneer Venus radio signals were received at multiple stations on Earth, allowing determination of both the east-west (zonal) and the north-south (poleward) components of velocity. The 90 m s⁻¹ zonal winds at cloud-top level are twice as fast as Earth's jet streams.

(Adapted from fig. 5 of Gierasch et al., 1997)[24]

the sound speed at those levels in the Venus atmosphere. The high speeds extend out to latitudes of ±45°, which is the extent of the Hadley cell on Venus. The first measurements of the zonal winds used cloud tracking from Earth-based telescopes and only sampled one altitude. Later measurements used the Doppler shift of radio signals from probes parachuting through the Venus atmosphere.[25] These measurements showed that the zonal winds decrease almost linearly with depth down to the surface where the wind speed is essentially zero. The zonal winds blow in the direction of the planet's rotation—from east to west, but 60 times faster than the solid planet. Hence they are part of the superrotation of the Venus atmosphere.

This superrotation is puzzling because one would expect the turbulent motions to slow it down. The friction force between layers of fluid sliding over one another is called viscosity. It arises from random molecular motions. However, molecular motions operate on microscopic scales, and it would take a very long time for molecular viscosity to slow the atmosphere down. Turbulent motions operate on macroscopic scales, which range up to planetary size for the largest waves and eddies. Thus turbulence has the potential to diffuse momentum much faster than random molecular motions and thus to slow down the atmosphere rapidly. The problem is that turbulence does not always act as a random diffusive process. The net effect of turbulence depends on many things—the speed and scale of the motion, how it is generated, and how it interacts with the rest of the flow. For example, some kinds of turbulence can transfer momentum into a zonal jet, tending to drive it away from a state of solid body rotation. This upgradient momentum transfer by turbulence has been called negative viscosity, although the term "viscosity" strictly applies only to motions on the molecular scale.

Some sort of upgradient momentum transfer seems to be necessary to maintain the superrotation of the Venus atmosphere. First consider the Hadley circulation, which involves rings of fluid that extend around the planet on lines of constant latitude. Because the superrotation exists at cloud top level at the equator, which is the widest part of the planet, the rings of fluid at the cloud tops at the equator have more angular momentum per unit mass than any other parts of the atmosphere. The direction of angular momentum around the north pole is clockwise, since that is the direction of atmospheric rotation. If the rings of fluid move away from the equator as part of the Hadley circulation, they will take their high angular momentum with them and they will be replaced by fluid with lower angular momentum. Thus the zonally symmetric Hadley circulation is draining angular momentum out of the equator, and an asymmetric (longitudinally varying) pattern of waves and eddies must be replacing it.[27]

An example of upgradient momentum transfer is a wave pattern that is tilted southwest to northeast (SW-NE) in the northern hemisphere and NW-SE in the southern hemisphere, like the swept-back feathers of an arrow pointing toward the west (box 3.14). The winds

Box 3.14.
Eddy Momentum Transport

The momentum transport can also be broken down into a symmetric part and an eddy part. The eastward momentum per unit mass is the eastward (zonal) velocity u, and the poleward mass flux is ρv (box 3.12), so the poleward transport of eastward momentum is $\int u \rho v dA$. This is similar to the expression for energy transport, except that momentum per unit mass u has replaced the dry static energy per unit mass $C_p + gz$. The breakdown into mean and eddy parts is as follows

$$\int u \rho v \, dA = \int \bar{u} \, \overline{\rho v} \, dA + \int \overline{u'(\rho v)'} \, dA$$

The overbar represents an average with respect to longitude. The first term on the right represents the mean zonal flow interacting with the symmetric overturning circulation. If the latter has poleward flow and large u at high altitude, and return flow and small u at low altitude, then zonal momentum will be transported poleward. The second term on the right is the eddy momentum transport. The primes are the fluctuations— the parts that vary in longitude after the mean quantities have been subtracted off. Eastward momentum will be transported northward if the eastward velocity component is correlated with the northward velocity component. In this case the flow lines would be tilted northeast-southwest (NE-SW). The opposite tilt, NW-SE, would correspond to westward momentum being transported northward. The direction of tilt is often observable and is an important diagnostic of the circulation.

blow parallel to the feathers, so a parcel moving south in the northern hemisphere is also moving west, and a parcel moving north is also moving east. Again we use the dump truck analogy. In the northern hemisphere, the southward moving parcels are carrying westward momentum, which they dump off at the equator. The dump trucks do double duty by taking eastward momentum out of the equator. Both effects accelerate the westward flow at the equator. The same transfer occurs in the southern hemisphere, although the tilt of the feathers is opposite. The net effect is to transfer westward momentum into the equator, where it is needed to maintain the westward superrotation.

Upgradient momentum transfer by tilted waves and eddies is observed in Earth's atmosphere and oceans, and in the atmospheres of Jupiter and Saturn. The eastward surface winds at midlatitudes of Earth are losing eastward momentum by friction with the surface. Yet the winds in the tropics are to the west, so simple diffusion would transfer westward momentum out of the tropics into higher latitudes, which is the opposite of what is needed to offset surface friction. Upgradient transfer by waves and eddies came to be known as negative viscosity when it was first observed on Earth. That term has fallen out of favor as people have come to understand it. The key is that the eddies have a separate source of energy, usually from buoyancy—hot air rising and cold air sinking, and when they find themselves in a shear flow like the zone between the low latitude flow to the west and the midlatitude flow to the east, they find themselves tilted like the feathers of an arrow. The tilting ultimately destroys the eddies by stretching them apart, but as they die they give up energy to the shear flow. The net result is that the shear flow is strengthened. Although the principle is accepted and the phenomenon has been observed on other planets, this upgradient momentum transfer has not been directly observed on Venus, largely because we can't see below the cloud tops.

There is no simple theory that accounts for the magnitude of the winds on all the planets. It is not a simple balance between the sources of energy like heat from the Sun and loss of energy by friction. The amount of potential energy at any one moment that could be converted into kinetic energy without further heating or cooling of the fluid parcels is called available potential energy (APE).[11,12,13] On Earth, and probably on Venus, most of the APE comes from the equator-to-pole temperature gradient. On Earth, sunlight is increasing the APE at a rate of 3.5% per day, and the large-scale circulation, including the Hadley cell, is reducing it at the same rate. On Earth the APE is about 4 times the kinetic energy (KE). The APE for Venus is not well known, and the rate of conversion is unknown. The relation between APE and KE on planets throughout the solar system is not well understood. Nevertheless, comparing the wind speeds and temperature gradients on Venus has the potential to increase our understanding of these basic climate variables on Earth.

4 MARS: LONG-TERM CLIMATE CHANGE

....

4.1 WARM AND WET OR COLD AND DRY

THE FASCINATION WITH MARS[6] HAS PARTLY TO DO WITH the possibility that life could have evolved there. Searching for evidence of liquid water, past and present, is therefore a major objective. Liquid water depends on the climate, so understanding Mars's climate, past and present, is also a major objective. As on Earth, climate change is recorded in Mars's sediments and ice deposits. The current climate is there for us to observe with telescopes, orbiters, and landers. There is much we don't understand about planetary climates, and studying Mars will help us understand climates in general.

Mars is currently a cold, dry place. At noon on the equator, which is the warmest place on the planet, the temperature soars to 20 °C. But there is no water at noon on the equator. Any thin layer of frost that might have condensed out of the atmosphere at night would have sublimed—vaporized without melting—into the dry atmosphere when the first rays of sunlight warmed the surface. The only permanent source of atmospheric water vapor is polar ice, which warms up to 205 K during summer and saturates the atmosphere at that temperature.

This is the dew point, the temperature below which the vapor in the air would start to condense and above which the frost on the ground would start to sublime. However, saturated air at 205 K contains very little water vapor, which never contributes more than a few hundred ppm (parts per million) of the molecules in the Martian atmosphere. For comparison, the dew point of saturated tropical air on Earth is 30 °C, and the water vapor makes up a few percent of the molecules.

Despite this bleak climate today, there is abundant geologic evidence that liquid water once flowed on the Martian surface. The evidence includes river channels, now dry, that were carved by powerful floods (fig. 4.1), valley networks of branching streams and tributaries indicating precipitation falling over wide areas (fig. 4.2), river deltas and layered sediments suggesting flow into standing bodies of liquid water, and salt and other minerals that dissolved in liquid water and were left behind when the water evaporated. Some scientists see ancient shorelines—evidence that an ocean covered the lowlands of the northern hemisphere to a depth of several hundred meters. When this wet period existed and when it ended is difficult to say.

The gold standard for dating a surface uses radioactivity. One collects rocks from the surface and measures their radioactive elements to see how much of the decay products have accumulated in the rock since it crystallized. For example, uranium decays to lead and potassium decays to argon, each at a known rate, so if the rock contains the parent and daughter elements, one can

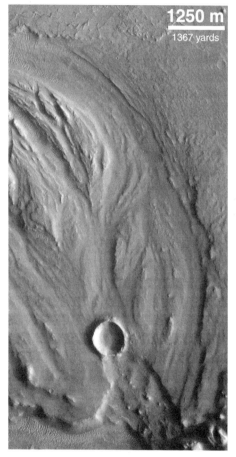

Figure 4.1. Mars flood channels. Geologic evidence suggests that the floods occurred during the first billion year of Mars's history. (Source: http://photojournal.jpl.nasa.gov/, PIA02076)

measure its age. Transferring this age to the age of a geologic unit—a widespread lava flow or sedimentary layer, can be tricky unless one knows where the rock came from. And one needs to collect the samples and bring them to a well-equipped laboratory.

The other way to date a planetary surface, which is used when rock samples are unavailable, is by counting craters. The longer a surface has been exposed to meteorite bombardment, the more craters it will have. This gives the relative age, and allows one to determine which surfaces are older and which surfaces are younger. To get the absolute age—to determine the age in years—one needs a standard. This was established for the Moon when lunar samples were returned to Earth as part of the Apollo program.[1,28] The crater counts of the areas of the Moon from which the samples came were carefully noted. The samples were dated in terrestrial laboratories using radioactive decay, and the relation between absolute age in years and crater density— numbers per unit area—was established. This relation is valid for the Moon, and it is possible that it is valid for Mars as well. Then counting craters on a Martian surface should yield its absolute age. It is also possible that the number of impacting objects is different for Mars. The problem has been carefully studied, and adjustments have been made, but the general conclusion is that the wet period of Mars's climate history occurred billions of years ago.

This causes a problem for Mars climate science, because the Sun was supposed to be fainter, perhaps by 20% or 30%, during the Mars wet period. The power output of the Sun—its luminosity over time—is based on models of its interior and evolution.[29] The models are well tested and agree with the luminosities of stars like the Sun whose ages are known. The same problem exists for Earth, which appears to have been warmer billions of years ago than it is today. Geologists have called this the "faint young Sun paradox." The solution to the paradox probably lies in climate science and not in the theory of stellar evolution, but the issue has not been entirely settled.

If Mars was wet billions of years ago, was it also warm? One could have had catastrophic floods released from underground aquifers where the top surface of the water froze like the top of a Hawaiian lava flow. The solid crust forms an insulating layer, trapping the heat of the liquid and allowing it to flow for great distances without freezing. Lava tubes on Hawaii are empty channels where lava once flowed in this way. Thus the climate of Mars could have been wet but not warm—the erosion could have been occurring entirely under the ice. Ice sheets and glaciers on Earth today have extensive plumbing systems that allow water to flow and then drain away. Liquid water at the base of glaciers acts as a temporary lubricant and allows them to surge and slide downhill. Flood channels, river deltas, and standing bodies of liquid water all could have existed on Mars under thick layers of ice. The wet but not warm hypothesis has the most trouble with the valley networks of branching streams and tributaries (fig. 4.2), which imply sources like precipitation over a wide area rather

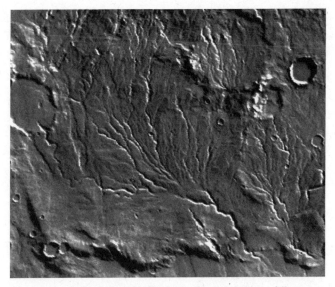

Figure 4.2. Mars valley network, which suggests that rain fell over a wide horizontal area. Alternatively, the water could have come from underground aquifers and flowed beneath an icy crust.
(Source:http://www.lpi.usra.edu/publications/slidesets/redplanet 2/slide_26.html)

than the bursting of an underground aquifer. Further study of Mars's geology will clarify these issues.

The alternate hypothesis, that Mars was wet and warm despite the faint young Sun, deserves study as well. The trick for both Earth and Mars is to accumulate enough greenhouse gases in the atmosphere to maintain a warm wet climate. Water vapor is not a significant contributor to the greenhouse effect on Mars today, because its abundance in the atmosphere is so low. Even at Earth orbit, the water vapor feedback is not capable of producing a regime change. The runaway greenhouse seems to occur between the orbits of Earth and Venus. This means that if one magically melted the polar ice caps of Mars and made the atmosphere temporarily warm and wet, it wouldn't stay that way. There's not enough sunlight to keep the planet warm, so after a period of intense rain and terrific floods, Mars would revert to its cold dry self. More to the point, the magical melting of the polar ice would never occur unless there was another more potent greenhouse gas to warm the planet. Could CO_2 fill that role?

Carbon dioxide is more potent than water vapor on a cold planet like Mars because it is more volatile—the pressure at which the vapor is in equilibrium with the frost is much higher for CO_2 than for water. For instance at 148 K, water ice is almost inert. Even under vacuum, it would sublime at a negligible rate. Its contribution to the mixture of gases in the Mars atmosphere would be tiny. Only when the ice warms up to 205 K does it make any contribution at all, and even then it is only a few hundred ppm, as discussed earlier. In contrast, CO_2 frost—dry ice at 148 K—sublimes until the CO_2 pressure is 6 mbar, at which point the gas is in equilibrium with the frost (box 4.1). This equilibrium probably governs the climate of Mars today. The balance between absorbed sunlight and emitted infrared at the pole, averaged over the year, sets the polar CO_2 frost temperature, which is 148 K. That temperature sets the pressure of CO_2 gas, which is the major constituent of the atmosphere.

Box 4.1.
Condensation and Evaporation

On Earth, air is saturated when the water vapor is in equilibrium with liquid or solid water—the condensed phase. Think of a wet fog where liquid droplets are suspended in the air, or a cirrus cloud where the suspended particles are ice crystals. In equilibrium, the rate at which the water molecules are striking and sticking to the solid or liquid surface is equal to the rate at which the surface is emitting water molecules. The rate depends on the pressure that the molecules of vapor are exerting on the surface, and in equilibrium, that pressure is called the saturation vapor pressure. For every substance, the SVP is a unique function of temperature $e(T)$. Mixing in other gases like N_2 and O_2 does not change the equilibrium condition. When other gases are present, the contribution of the water vapor to the total pressure is called the partial pressure, and equilibrium is when the partial pressure of the vapor is equal to the SVP of the solid or liquid.

Evaporation occurs when the SVP is greater than the partial pressure. Evaporation from ice is called sublimation. Evaporation takes energy—the evaporating water absorbs heat from the system, which is how we cool off by sweating. Condensation occurs when the SVP is less than the partial pressure. Condensation releases heat, which is how a glass of ice water warms up on a summer day. The energy absorbed or released is called latent heat, and the latent heat per unit mass of vapor L is a unique function of temperature for each condensing substance.

Transfer of latent heat has a stabilizing effect. It tries to bring the components of the system into equilibrium. Consider a vapor and a condensed phase—either solid or liquid—in a closed container. If the SVP of the condensed phase is lower

(*Box 4.1 continued*)

than the partial pressure of the vapor, the resulting condensation will warm the condensed phase and raise its SVP. Also, the condensation will remove some of the vapor and lower its partial pressure. Both effects tend to bring the two pressures closer together. If another gas like air is present, it may act as a barrier and slow the relaxation to equilibrium. However in a pure system, like dry ice in a CO_2 atmosphere, the equilibrium is almost instantaneous.

The only place that other gases enter is in boiling, which depends on the pressure of the environment and is not a unique function of the condensing substance. Boiling occurs when the SVP of the condensing substance is equal to the total atmospheric pressure, including the contributions of all constituents. Then, a bubble of pure vapor can withstand the outside pressure, which means that vapor bubbles can spontaneously erupt inside the liquid and cause it to boil.

4.2 DRY ICE AND WARM CO_2 ATMOSPHERES

The model that posits a warm and wet early Mars goes something like this: Mars once had large polar deposits of solid CO_2, whose temperature determined the pressure of CO_2 in the atmosphere.[30] The amount had to be many times the mass of the current atmosphere, enough to make a huge greenhouse effect and melt the polar ice caps. The amount, expressed in terms of surface pressure, is 1 bar or more,[31] not on a par with Venus but certainly much more than the CO_2 in Earth's atmosphere. The 1 bar amount depends on the assumptions of the model and is highly uncertain. It is even possible that no amount of CO_2, by itself, could produce a warm, wet Mars. The problem is that a massive CO_2 atmosphere raises the planet's albedo and thereby limits the temperature rise.[32]

Nevertheless, let us take 1 bar as the minimum value. Since the current surface pressure on Mars is 6 mbar, the mass of this CO_2 reservoir has to be more than 100 times the mass of the current Mars atmosphere. Then let Mars tilt its polar axis toward the Sun. Such changes in the tilt angle—the obliquity—are well established and are part of the known astronomical cycles that cause the Earth's spin axis to change. The astronomical cycles, also known as Milankovitch cycles, contribute to climate change on Earth. The difference is that the amplitudes of the changes are much larger for Mars than for Earth.[33] Mars gets pushed around by the other planets much more than Earth does, and its tilt angle swings up and down over millions of years between 15° and 45°. During periods of high obliquity the polar caps would warm up, and there would be a massive outpouring of CO_2 gas into the atmosphere. The greenhouse effect of all that gas would warm the planet further, and at some point water vapor would join the fray, causing even further warming. This is similar to the runaway greenhouse effect, but it is initiated by CO_2, which is more volatile than water and therefore would have been more able to respond on a distant planet like Mars when the Sun was young and faint. The wild swings of obliquity are essential to this process, and they are fairly certain. The massive reservoir of CO_2 is the other essential element. If the reservoir never existed, the warming by CO_2 never could have happened.

Is there any evidence that Mars once had a massive CO_2 atmosphere? The atmosphere of Mars today is 95.5% CO_2, 2.7% N_2, and 1.6 % Ar, with CO, O_2, H_2O, and noble gases making up the remainder (table 2.1).[17] The mass of atmosphere per unit area is 1.6% of the mass per unit area of Earth's atmosphere. If Mars once had a CO_2 atmosphere 100 times more massive than the present atmosphere, then the CO_2 either is buried somewhere on Mars or else it escaped to space. If the climate cooled and the massive CO_2 atmosphere became the present polar caps, the question is, why did the climate cool when the Sun was heating up? Equivalently, what sustained the greenhouse then, when the Sun was fainter? A further complication is that the current polar caps seem to be mostly water ice, not CO_2. Although a meter of CO_2 frost accumulates during the fall and winter, it sublimes during the spring and exposes water ice during the summer. Orbiting spacecraft have probed the ice with penetrating radar and have shown the ice is more than 3 km thick in places. From the propagation speed through the ice, the radars suggest that it is mostly water. Pockets of CO_2 frost may be present, but their total mass is less than that of the current atmosphere.[34] The implication is that any 1-bar atmosphere of the past cannot be stored as CO_2 frost today.

Carbonates are another place to store a 1-bar CO_2 atmosphere. If Mars was once warm and wet with standing bodies of liquid water, the weathering of silicate rocks and precipitation of limestone could have turned atmospheric carbon dioxide into carbonate rocks, as they have done on Earth. If uniformly spread over the planet, the limestone layer would have to be tens of meters thick, and no such layer has been detected. There are some carbonates exposed on the surface,[35] and they may have formed in liquid water, but they cannot account for the massive CO_2 atmosphere needed to produce the liquid water in the first place.

The other possibility is that the CO_2 escaped to space. Mars gravity is significantly weaker than that of Earth and Venus, so escape processes are more effective. Sputtering, where molecules in the upper atmosphere become electrically charged and are then carried away by the solar wind, is more effective than on Earth and Venus. Photochemical escape, where a molecule absorbs a photon and then splits apart—dissociates—into fast-moving fragments is also more effective. The high D/H ratio (fig. 2.2) is evidence of hydrogen escape and loss of water. The small abundance of nitrogen and the noble gases, compared to their abundances on Earth and Venus, is evidence that these gases have escaped. The nitrogen isotope ratio $^{15}N/^{14}N$ is 1.6 times that of the Earth.[36] That ratio shows evidence of escape, but it is less than what one would expect given the low total abundance of nitrogen. The isotope ratios of carbon ($^{13}C/^{12}C$) and oxygen ($^{18}O/^{17}O/^{16}O$) are very similar to those on Earth, which seems inconsistent with massive escape of CO_2. The history of volatile compounds on Mars is an area of active study.

4.3 TERRAFORMING

Could humans change the climate of Mars enough to make it habitable? The first step would be to make it warm and wet. Then one could set about altering the atmospheric composition, perhaps by planting crops that turn water and carbon dioxide into hydrocarbons and oxygen. One would want to raise the atmospheric pressure, too. The subject is too broad and speculative for this book, but we will discuss the first step. We already mentioned some of the difficulties in trying to understand a warm wet climate billions of years ago. The faint young Sun makes those climates less likely than a warm wet climate today. On the other hand, there was probably more water and carbon dioxide on Mars during the first two billion years, because those substances may have escaped during the intervening time. So let us start by asking, how much water and CO_2 does Mars have today?

If water is abundant, one expects to find permafrost in the polar regions. Given the partial pressure of water vapor in the atmosphere (box 4.1), one can calculate the latitude boundary of the permafrost. Closer to the equator, the frost will be too warm, and it will evaporate, slowly diffusing upward through the soil and into the atmosphere. There it will circulate around and finally be trapped in the coldest regions on the planet—the polar regions. These calculations were tested by the gamma ray spectrometer (GRS) on the Mars Odyssey spacecraft.[37] The GRS detects the chemical elements that are in the atmosphere and in the soil down to a depth of

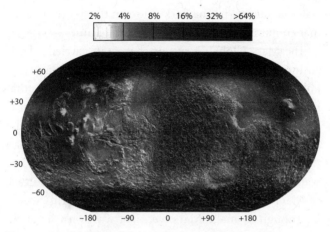

Figure 4.3. Lower limit of water mass fraction on Mars. The instrument detects hydrogen atoms in the upper meter of crust, and the hydrogen is interpreted as a sign of water. Latitude and longitude are given in degrees. In the polar regions, the concentration of water ice is in the range 35–50% by mass.

(From fig. 6 of Boynton et al., 2002.[37] Additional source: http://photojournal.jpl.nasa.gov/, PIA04907)

about one meter. An element shows up no matter which compound it is associated with. Hydrogen is most likely present as water, so its presence is taken as a detection of water. Figure 4.3 is a GRS map of Mars showing that water ice is present in the soil down to latitudes of 45° in both the northern and southern hemispheres. Even patches near the equator contain water, which may take the form of ice under a thick insulating blanket of dust or else chemically bound water in clay minerals. Since the GRS probes down only to 1 m depth, the total amount of water is unknown. The dark colors in the figure are soils that contain at least 50% water ice by volume.

The Mars Global Surveyor spacecraft carried a laser altimeter that mapped the topography in exquisite detail. At each of the poles there is a topographic bulge that rises ~3 km above the surrounding plain.[38] Each bulge is a stack of layers that are exposed around the edges. The combined volume of the two bulges is a little less than the Greenland ice sheet. If these topographic bulges are water ice, and were to melt, they would cover the surface of Mars to an average depth of ~19 meters. This would be enough to make Mars wet, if it was warm enough. The topographic bulge in the north is covered by water ice, and the layers are water ice with various amounts of dust. Sunlight reflected off the surface of the polar cap has the spectral signature of water ice, but the measurement probes only the upper few millimeters. The surface temperature rises to 205 K in the summer, after the winter deposit of CO_2 frost has sublimed away.[39] Such temperatures are what one expects of water ice and are much too high for CO_2, which cannot warm above 148 K in a 6-mbar atmosphere (box 4.1).

The south polar layered deposits are more complicated. The exposed residual frost—the frost that survives the summer—covers a smaller area than the residual frost in the north, and it does not fully cover the topographic bulge. The exposed layered deposits are a mixture of dust and ice. The visual color is that of dust, but the infrared spectrum contains the signature of water ice as well. Moreover, the residual frost contains both water ice and CO_2 ice. The CO_2 part is a flat-topped upper layer that appears to have collapsed in various places, forming quasi-circular pits 100 to 1000 meters in diameter.[40] This flat surface with pits has been called the Swiss cheese terrain (fig. 4.4). The CO_2 layer is ~8 m thick. The bottoms of the pits are flat, but they have their own thin layers in the middle. The deepest parts of the pits are around the edges and are called moats, like the trenches around a medieval castle. The high, flat-topped CO_2 areas are called mesas. The bottoms of the moats are water ice, and they warm up in the summer to temperatures well above the CO_2 equilibrium temperature.[41] The mesa tops stay at the equilibrium temperature for the entire year. The curious thing is that the diameters of the pits are growing at the rate of 4 meters per Earth year. This confirms that the upper layer is CO_2, since water ice is much less volatile at Mars's polar temperatures. The growth reflects collapse of the walls and sublimation of CO_2 frost. But the rapid growth implies that the upper layer will be gone in 100–200 years, when overlapping pits cover the entire area. Overlapping pits encroaching on the mesas are visible in figure 4.4.

The disappearing CO_2 layer is a puzzle, for several reasons. First, why should humans be exploring Mars during this special hundred-year time interval? Second, where is the rest of the frozen CO_2 that is supposed to control the partial pressure of CO_2 in the atmosphere? If the CO_2 layer is only 8 m thick, then its total mass is only 3.5% of the total atmospheric mass.[42] If the obliquity were to increase, even by a small amount, the CO_2

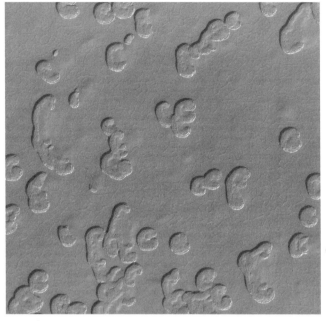

Figure 4.4. Mars Swiss cheese terrain. These quasi-circular depressions are found only on the south polar ice cap. They are hundreds of meters across, 8–10 meters deep, and appear to be growing in diameter at rates of 3–4 m per Earth year.

(Source: http://www.msss.com/mars_images/moc/nature_polar/np_2a/)

frost would evaporate completely. Why should we be exploring during this special time, when the obliquity is so finely tuned to the CO_2 cycle? Finally, if the atmospheric mass and the mesas are the entire reservoir of CO_2, why is CO_2 so rare on Mars and so abundant on Earth and Venus? The latter could be due to massive escape of CO_2, but the isotopes of carbon and oxygen do not provide clear evidence of escape. In any case, if almost all the CO_2 is already in the atmosphere, the prospects for starting a greenhouse effect by vaporizing massive amounts of CO_2 are close to zero.

Of course, the moat spectra are only skin deep—they tell us that water ice is present 8 meters below the surface of the CO_2 layer, but they do not tell us what lies below. Here radar sounders on two different spacecraft, one from the United States and one from Europe, have made the difference. The radio waves can penetrate through kilometers of ice, and have detected the bottom of the layered deposits at depths up to 3.3 kilometers. Making the assumption that the bottom layers are flat, the scientists find that the time for the radar pulse to propagate from the spacecraft to the bottom layer and back is generally consistent with the speed of radio waves in water ice. However, in some places the speed is more consistent with CO_2 ice. The pockets of CO_2 ice are up to 600 meters thick, and they contain almost as much mass as the atmosphere itself.[37] This reduces the coincidence of living in special times, but it still does not allow us to terraform Mars. Even doubling the amount of atmospheric CO_2 would not start a greenhouse effect powerful enough to melt the water ice and bring on the added greenhouse effects of water vapor. Earthlings may have to wait for the solar luminosity to increase over the next few billion years before the climate of Mars becomes truly Earth-like.

5 MARS: THE PRESENT ERA

....

5.1 SEASONAL CYCLES

MARS HAS SEASONAL CYCLES, AS DOES EARTH. ITS obliquity is 25° and the Earth's is 23°. At the poles of each, frost accumulates during the fall and winter and evaporates during the spring and summer; and right at the poles, the frost lasts throughout the year. In these respects the seasons on Mars are like the seasons on Earth, but there are differences. First, on Mars, there are two kinds of frost—water and CO_2. Second, the frosts never melt—both substances go directly from the solid phase to the vapor phase and back without an intermediate liquid phase. Third, one of the condensing substances, CO_2, is also the major constituent of the atmosphere. Since the atmospheric pressure is less than 1% of the Earth's atmospheric pressure, the build-up of CO_2 frost at the winter pole causes a 20–30% drop in atmospheric pressure. This drop occurs twice per year, once during northern winter and once during southern winter. The eccentricity of the Mars's orbit makes the southern drop deeper than the northern one.

Several key concepts are necessary in discussing evaporation and condensation (box 4.1). One is saturation vapor pressure (SVP), which is the pressure of the vapor when it is in equilibrium with a condensed phase, either solid or liquid. For every substance, the SVP is a unique function of temperature. Mixing in other gases like N_2 and O_2 does not change the equilibrium condition. When other gases are present, the contribution of the water vapor to the total pressure is called the partial pressure, and equilibrium occurs when the partial pressure of the vapor is equal to the SVP of the solid or liquid. The SVP is a strong function of temperature (box 5.1). The SVP of water is 6.1 mbar at $T = 273$ K, and it approximately doubles for every 10 K increase of temperature.

Evaporation occurs when the SVP is greater than the partial pressure (box 4.1). Condensation occurs when the SVP is less. Evaporation from ice is called sublimation. Evaporation absorbs heat from the system, and condensation releases it. The energy absorbed or released is called latent heat, and the energy per unit mass of vapor is a unique function of temperature for each condensing substance. At low temperatures, the latent heat is essentially constant. Absorption and release of latent heat tend to bring the vapor and the condensed phase into equilibrium with each other because condensation warms the condensed phase and increases its vapor pressure. Also, in a closed system at least, condensation removes vapor and reduces its partial pressure. Evaporation is the reverse, but in both cases the SVP and the partial pressure approach each other and come closer to equilibrium. For a pure vapor like CO_2 in Mars's atmosphere, the approach to equilibrium with the condensed phase is almost instantaneous. Thus solid CO_2 at an average elevation on

---

Box 5.1.

Clausius-Clapeyron Equation

For the condensable gases in planetary atmospheres, including CO_2, CH_4, NH_3, and water, the SVP is a strong function of temperature. If the density of the gas is small enough it will obey the ideal gas law and an approximate form of the Clausius Clapeyron equation:

$$e(T) = \rho R_g T, \quad \frac{1}{e}\frac{de}{dT} = \frac{L}{R_g T^2}, \quad e(T) \propto \exp\left(\frac{-L}{R_g T}\right)$$

The last step follows if one assumes L is constant, which is a good approximation at low temperatures. For water near 0°, the value of $L/(R_g T)$ is ~20, and $e(T)$ approximately doubles for each 10° increase in temperature. At 0 °C, 10 °C, 20 °C, and 30 °C, the SVP in mbar is 6, 12, 23, and 42, respectively, so the percentage of water molecules in saturated air at 1000 mbar is 0.6, 1.2, 2.3, and 4.2, respectively. CO_2 is much more volatile than water—its SVP is much higher than that of water at the same temperature, and its temperature at a given SVP is much lower. For water, the temperature at SVP = 1 bar is 100 °C. This is the boiling point of water at sea level. For CO_2 ice, the temperature at SVP = 1 bar is −78 °C. The temperature at SVP = 6 mbar, which is the average surface pressure on Mars, is −125 °C or 148 K. Since the atmosphere is almost pure CO_2, the atmosphere and frost are always in equilibrium. So 148 K is the temperature of all solid CO_2 at the surface of Mars.

Mars always has an SVP of 6 mbar, for which the equilibrium temperature is 148 K (box 4.1). At higher elevations, where the atmospheric pressure is less, the equilibrium temperature is a few degrees less as well.

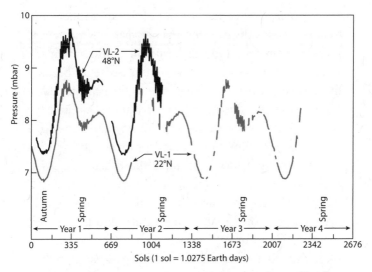

Figure 5.1. Pressure variations over 3 Mars years as observed by the Viking Landers. The x-axis is time in Martian days. Lander 1 was at a higher elevation than Lander 2 and so experienced lower pressures. The semi-annual drops in pressure occur when CO_2 frost accumulates at each pole in winter.

(Source: http://www-k12.atmos.washington.edu/k12/resources/ mars_data-information/3.3years.gif. Adapted from Fig. 1 of Tillman et al., 1993)[43]

Figure 5.1 shows the pressures measured by the two Viking landers (VL1 and VL2) over the course of three Martian years[43]. VL2 was at a lower elevation than VL1, so the pressures were uniformly higher at the VL2 site. The two landers were almost 180° apart in longitude, and despite the great distance between them, the pressure curves march up and down with the seasons at almost exactly the same rate. During each winter season at each pole, CO_2 condenses out of the atmosphere and frost builds up. The two minima per year in atmospheric pressure correspond to northern and southern winters, respectively. The southern minimum is deeper because the southern winter is longer, due to the eccentricity of Mars's orbit. As a result, more frost builds up and the atmospheric pressure dips to a lower value. The frost at each pole sublimes during its spring season, and the atmospheric pressure rises until the other pole starts accumulating frost during its fall season. Then the atmospheric pressure starts to fall, and the semi-annual cycle repeats. The broad-scale maxima and minima are reproducible from year to year, indicating that the seasonal growth and retreat of frost is regular and predictable. Although the sublimation and condensation are confined to the polar regions, the pressure of the atmosphere goes up and down simultaneously at all locations. This is because the atmosphere cannot tolerate a vacuum or a pressure excess: As gas is removed or replaced at one pole, more gas immediately flows in to take its place, and the whole atmosphere adjusts as a unit. This breathing up and down occurs because CO_2 is the major constituent of the atmosphere.

5.2 WATER

The behavior of water today may be a clue to the behavior of water in the past, when Mars was wet and possibly warm. Figure 5.2 shows the seasonal cycle of atmospheric water vapor on Mars.[44] Water behaves differently

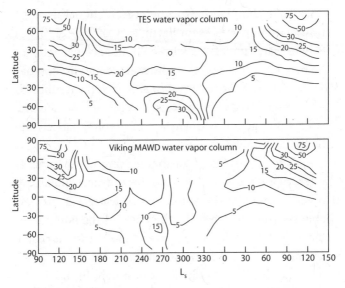

Figure 5.2. Viking water vapor vs. latitude and season. The x-axis is L_s—the angular position in degrees of Mars in its orbit around the Sun, with $L_s = 0$ on the first day of northern spring. The contours are labeled in units of precipitable microns—the depth of the liquid layer that would result if all the water vapor in the atmosphere were to condense. Each point is a zonal average of the water vapor at that latitude and season. The largest water vapor abundances occur over the north polar cap in summer.

(Adapted from fig. 5 of Smith, 2002)[44]

from CO_2 because water vapor is a minor atmospheric constituent—less than 0.1% of the molecules are water whereas 95.5% are CO_2. Removal or addition of water molecules from the atmosphere does not create a significant vacuum or pressure excess, and the atmosphere does not move to correct the irregularities. Therefore the water vapor abundance varies by large fractional amounts from place to place. As on Earth, each air parcel carries its own water molecules whose abundance depends on the history of that parcel—how much condensation has occurred to remove water vapor and how much sublimation has occurred to put it back. The history includes the parcel's interaction with surface reservoirs, particularly the ice in the polar regions.

As shown in Fig. 5.2, the maximum amount of water vapor each year is near the northern polar cap in early summer. The amount is measured in precipitable microns, which is the thickness of the liquid layer that would form if all the water vapor in the air were condensed out. The maximum amount is about 75 microns—0.075 mm, which is not much. In the Earth's tropics, the layer would be tens of cm thick. If 75 microns of water were mixed uniformly with CO_2 up to the top of the Martian atmosphere, then the abundance of water molecules would be 100 ppm. Equivalently, the contribution of water molecules to the Martian atmospheric pressure—the partial pressure—would be 6×10^{-4} mbar, which is small compared to the 6-mbar total pressure. In fact the water vapor is concentrated closer to the ground because the atmosphere is warmer there, so the partial pressure near the ground over the north polar cap at the beginning of summer is 2–3 times higher than the 6×10^{-4} mbar value, perhaps in the range $1.5–2.0 \times 10^{-3}$ mbar. This pressure is equal to the SVP of water ice at ~205 K, which is the temperature of the north polar cap in summer.

In late spring the water ice in the northern polar cap is still covered by CO_2 frost,[39] which is in contact with a 6-mbar atmosphere of CO_2 and is therefore at a temperature of 148 K. This frost sublimes away during spring and is gone by the beginning of summer, at which point the temperature of the water frost shoots up to 205 K. Water is not liquid at this temperature, and the ice does not melt. The temperature reflects a balance between absorbed sunlight and emitted infrared radiation. As the frost warms to 205 K, its SVP increases to $1.5–2.0 \times 10^{-3}$ mbar, which is initially much higher than the partial pressure of water vapor in the air. This imbalance causes rapid sublimation until the air above the north polar cap is saturated with ~75 precipitable microns of water in the atmospheric column. This period of high (for Mars) water-vapor content doesn't last long, however, for two reasons: First, the winds are constantly blowing the saturated air away from the polar cap, and second, the Sun moves lower in the sky later in the summer, and the temperature of the water frost decreases. Figure 5.2 shows a tongue of water vapor, represented by the 20 precipitable micron contour, making its way down to the equator by the end of summer, but at that point the air is much dryer, having lost much of its water to condensation.

Condensation occurs on solid surfaces, either the ground, the cloud particles, or the dust particles in the atmosphere. Dust particles are always present on Mars. They are micron-sized (10^{-6} m) mineral grains that are picked up by the wind and take months to settle out once they are in the air. Like cloud particles, their temperature is the same as that of the air around them, so the cloud and dust particles will acquire water and grow when the air is sufficiently cold—when the air is saturated. If the cloud particles fall out—precipitate—then the air loses its water. Frost forms when the ground is sufficiently cold—when the air in contact with the ground is saturated. On Mars the maximum partial pressure of water is the SVP of the warmest exposed ice deposits, which are at the north polar cap in early summer.

There is much uncertainty about the possibility of liquid water at the surface of Mars today. With the low atmospheric pressure, liquid water tends to boil even when it is close to freezing.[45] Boiling occurs when the SVP of the liquid exceeds the ambient pressure. The SVP at 0 °C is 6.1 mbar, which is close to the average surface pressure on Mars. At elevations above the average, even ice water would boil. Vapor bubbles would form in the liquid, thereby increasing its surface area and hastening evaporation. The buoyancy of water vapor (18 g mole^{-1}) in a CO_2 atmosphere (44 g mole^{-1}) further hastens the evaporation. Evaporation cools the liquid. Under average Martian conditions, any liquid water set out on the surface would quickly freeze. The energy in sunlight is not enough to offset the cooling effect of evaporation, but there are ways liquid might exist nonetheless. One way is to seal it under the ice, so that the liquid can't evaporate. Another way is to have it full of dissolved salts, which act as antifreeze and lower the freezing point. A third way is to have the liquid in monolayers, where binding to the rock keeps the water liquid even at low temperatures.

These options reduce the biological potential, since they lower the chemical activity of the water. The best observational evidence for liquid water today is the existence of the gullies on the insides of craters.[46] They appear to start just below the rim, usually from a single layer. Whether they are landslides, dust avalanches, or seeping water is an open question. The liquid might exist in an underground aquifer and then freeze over when it is exposed to the Martian atmosphere. It could still carve the gullies by flowing downhill underneath the ice.

The Viking landers and subsequent landers have photographed thin layers of frost on the ground. For frost to form, the temperature of the ground must fall below the dew point, which is 205 K at the seasonal peak. The lowest temperatures occur just before dawn. The layers are thin because the atmosphere holds no more than ~75 precipitable microns, and not all of that precipitates out. Because they are thin, the layers disappear in less than an hour once the sunlight touches them. Ice clouds are common as well. Telescopes on the Viking orbiters, telescopes on Earth, and the Hubble Space Telescope all have imaged thin water ice clouds in the Martian atmosphere. It is not known whether the ice particles fall out on the ground or evaporate before hitting the ground. The latter case merely redistributes the water and does not necessarily remove it from the atmosphere. Eventually, during the fall and winter seasons, the water that was released from the north polar cap during the beginning of summer finds its way back, where it is cold-trapped by the CO_2 frost that is being deposited at 148 K. There may be a small net transfer of water into the soil and a small net transfer to the south polar region, but such transfer becomes significant only over tens of thousands of years.

A major question is why the north polar cap provides more water to the atmosphere than the south polar cap. Figure 5.1 shows that the south polar cap acquires a greater covering of CO_2 frost during its fall and winter season than the north polar cap does. This is due to the eccentricity of the orbit: Mars is farther from the Sun and it moves around more slowly in its orbit during southern fall and winter, so there is more time for the frost to accumulate. On the other hand, Mars is closest to the Sun during southern spring and summer, so the sunlight is more intense and the seasonal frost covering should sublime faster. In principle, the effects of orbital eccentricity should cancel out, but they don't. The southern cap stays colder during its summer than the northern cap, and it releases less water vapor to the atmosphere. Also, it has CO_2 frost deposits that survive the summer, and the northern cap does not. This is an unexplained difference between the two hemispheres. One possibility is that the Martian climate is not in a steady state—that the current configuration of the polar ice deposits is a remnant of some past climate that is still adjusting to a change in climate forcing. Milankovitch cycles are the obvious forcing functions, and they do change drastically over periods of millions of years. Moreover, water moves around slowly on Mars, because there is so little of it in the atmosphere at any one time. If the north polar cap lost 75 precipitable microns of water every Martian year, it would take twenty-five million Mars years to move an ice deposit 3 km thick from one hemisphere to the other. Even if the annual loss were twenty-five times greater than the 75 microns that are in the atmosphere at any one time, it would still take a million years for the climate to fully respond to the Milankovitch cycles. Climate change happens slowly on Mars because the planet is so cold and dry.

5.3 WEATHER AND WIND

In the Viking lander data, the main differences from one year to the next are associated with the weather.[47] In figure 5.1, weather appears as high-frequency oscillations, which seem to be greatest during northern fall and winter, especially at the VL2 site. The amplitude of the oscillations is not as large as the seasonal swings associated with frost growth and retreat, and is less reproducible from one year to the next. The period of the weather oscillations ranges from 2 sols to 8 sols,[48] where 1 sol = 1 Martian day = 1.027 Earth days. In many respects, the weather on Mars is like the weather on Earth. First, the weather disturbances are largest during the fall and winter seasons. The VL2 site is at latitude 48°N, and the VL1 site is at latitude 22°N. Thus both sites are in the northern hemisphere, and both experience winter storms, but the storms are more intense at the higher-latitude site. Second, the period of the storms is similar to the period of winter storms on Earth. These are the midlatitude cyclones and anticyclones and their associated warm fronts and cold fronts that move from west to east and bring us our winter weather. The 2–8 day time interval is a combination of two effects: One is the typical lifetime of individual storms, and the other is the rate at which they move past us. Both effects conspire to make it difficult to forecast the weather more than a week ahead. As on Earth, the winter storms draw their energy from the temperature contrast between the cold winter pole and the warm equator. The temperature contrast is much less in the summer hemisphere, since the Sun warms the pole and equator almost equally in the summer.

One difference between Mars weather and Earth weather is the importance of the diurnal cycle—the swings of temperature and wind that accompany the rising and setting of the Sun.[47] Our atmosphere has enough mass to retain most of its energy at night and absorb most of the solar energy during the day without a huge change in temperature. Most of the diurnal changes occur in the atmospheric boundary layer, within 500 meters of the surface. Often the boundary layer is even thinner. On cloudless winter nights, when the temperature of the ground drops below freezing, orange growers use motor-driven propellers to stir the boundary layer. This brings air down from above, where the temperature is less affected by the diurnal cycle and is therefore warm enough to keep the crop from freezing. The Mars atmosphere has much less mass than Earth's, so the diurnal swings are larger and extend to higher altitudes. Daytime temperatures at the equator rise above 0 °C, and at night the temperatures fall below −100 °C.

The wind speed and direction vary in regular ways throughout the day on Mars. Two processes are at work. One is the global flow pattern in response to the bulge of warm air on the side of Mars that faces the Sun. Since Mars rotates from west to east, the bulge sweeps across the surface of Mars following the Sun from east to west. At the VL2 site, the wind direction rotates clockwise, completing a full circle every day. The global flow pattern is called a thermal tide, to distinguish it from a gravitational tide. Both are responses to the effects of another celestial body—in this case, the Sun. A thermal tide is a response to the Sun's heating, and a gravitational tide is a response to the Sun's gravitational attraction. The other process that has a strong diurnal component is upslope-downslope motion. These types of winds are common in mountainous areas on Earth. The Sun warms the slopes during the day, and the air close to the ground rises. Infrared radiation cools the slopes during the night, and the air close to the ground sinks. The downslope winds at the VL1 site reached speeds of 6 meters per second, peaking out shortly before midnight on every sol. The atmospheric boundary layer is thinner during the night than during the day because the cold ground stabilizes the air. Therefore the evening winds are concentrated closer to the surface, which makes them stronger than the daytime winds. During the day the boundary layer is warmer than the air above, and it spreads upward by convection, which dilutes the intensity of the winds.

In the spectrum of planetary climates, Mars lies off in the direction of planets with no atmosphere at all. But it's not quite there; the Moon is a better example. On the Moon the diurnal swings of temperature are huge— it's hot enough to fry an egg at noon and almost cold enough to liquefy nitrogen at night. Without an atmosphere there are no winds, but with a thin atmosphere the winds would be slave to the diurnal cycle of temperature. The weather would be entirely predictable—the time of day would tell you which way the wind is blowing. The mass per unit area of the Mars atmosphere is almost 2% that of the Earth, so Mars is not as extreme as the Moon. Nevertheless, the weather is usually more predictable than Earth's weather. The diurnal cycle is relatively more important, and the winter storms are relatively less important. In that respect, the weather on Mars is boring.

5.4 DUST STORMS

Dust storms are the exception to the boring weather of Mars. Dust devils are like small tornadoes, and global dust storms envelop the planet for 30–60 days. The global storms tend to occur during the southern spring, when Mars is closest to the Sun, but they do not occur every year. This makes them difficult to forecast, which is a problem because they have a major impact on conditions at the surface.

Since Mars is so dry, there is no moisture to hold the dust down. There is no freeze-thaw cycle, so the ice doesn't bind the grains together. Instead, the winds are constantly stirring the dust. The dust storms range in size from dust devils 10 meters in diameter to global dust

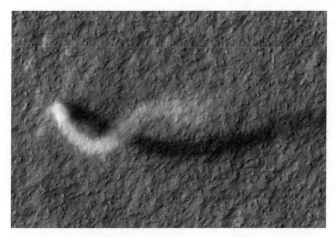

Figure 5.3. Mars dust devil. The length of the shadow indicates that the dust devil's height is at least 800 m. Its width is 30 m. Some dust devils extend to heights of 8–10 km.
   (Source: http://photojournal.jpl.nasa.gov/, PIA15116)

storms that cover the planet. Dust devils are common. They resemble tornadoes, but their winds are weaker. Some rise to heights of 8 km. Rovers on the surface have made time-lapse photos of dust devils drifting with the prevailing wind. Spacecraft orbiting the planet have photographed dust devils rising 10 km into the atmosphere (fig. 5.3). The dust devils leave dark tracks where they have blown the dust away and exposed a dark substrate. An unplanned benefit of dust devils is that they clean the solar panels of Mars rovers and landers. The solar panels turn sunlight into electricity and power the vehicles. The constant fallout of dust from the atmosphere gradually covers the solar panels and would eventually cause the vehicles to fail. Instead, a dust devil usually comes through and blows the dust away. Dust devils are partly responsible for the dust, but they help clean it up. Although individual dust devils are unpredictable, collectively they can be counted on to maintain the atmospheric dust at a certain level. Even on a clear day, the optical depth τ on Mars is about 0.3, which means that roughly 30% of the incident sunlight is either scattered or absorbed before reaching the ground.

Global dust storms develop in certain favored regions, which are presumably the places where the wind is strongest. Once the dust is aloft, it absorbs sunlight and heats the air, causing it to rise. The winds near the surface—in the atmospheric boundary layer—converge on the site where the air is rising and bring in more dust. The temperature difference between the heated air inside the dust storm and the air outside helps to amplify the wind. The temperature rise during a global dust storm can be as much as 30 °C. A global dust storm shuts down when the suspended dust is spread uniformly over the planet. Although the temperature is higher, the thermal contrast is less and the winds decrease. It still takes a month or more for the dust to settle out of the atmosphere. Viewed from space, surface features disappear during a global dust storm. Although the effect on atmospheric temperatures and visibility is large, the amount of dust is small—it forms a layer of order 10 microns deep once it has settled out.

Global dust storms are difficult to predict. They occur when Mars is closest to the Sun, which happens when the season is late southern spring, but they don't occur every year. In that sense they are like the El Niño, which occurs around Christmas time but doesn't occur every year. We blame the ocean for this interannual variability. The warm water in the upper 100 meters of the tropical Pacific Ocean sloshes back and forth from the western Pacific to the eastern Pacific. When the warm water is over on the eastern side, the Peruvian fisheries suffer because the warm water acts as a cap that prevents the colder nutrient-laden waters from welling up to the surface and feeding the local food chain. The warm water off the west coast moisturizes the atmosphere and increases the winter rainfall in the southwestern United States. Although we cannot predict the El Niño many years ahead, we can count on it to persist for a season or two. The persistence has to do with the mass of the ocean and its slow modes of variability. Why some years have an El Niño and others don't has to do with the long-term variability of the oceans.

The problem is, Mars doesn't have oceans but it does have interannual variability. The Mars atmosphere has less than 2% of the mass per unit area of Earth's atmosphere. The persistence of weather patterns should be much shorter on Mars than on Earth, and on Earth weather patterns persist for only a week. Except for the day-to-day fluctuations of the weather, what causes one Martian year to be different from the next? One possibility is that dust on the ground provides the inertia to the Mars climate system.[49] Global dust storms may clean out the windy places, so there is no dust for the winds to pick up. According to this hypothesis, it takes several years for atmospheric fallout to replenish the dust in the windy places. Once the resupply is complete, the system is ready for another global dust storm. This hypothesis says that the gap between one global dust storm and the next is related to the time needed for the dust to accumulate. If the hypothesis is correct, one could forecast the next global dust storm by measuring the depth of dust in the places where the wind blows hardest.

6 TITAN, MOONS, AND SMALL PLANETS

6.1 ATMOSPHERIC EVOLUTION

TITAN[9] IS A MOON OF SATURN THAT ACTS LIKE A TER-restrial planet. It's fairly successful, too, but it's had to use different raw materials to do it. The basic problem is the low temperature. Saturn and Titan are ten times farther from the Sun than Earth is, so the energy in sunlight is just 1% of the terrestrial value. Thus Titan's surface temperature is −179 °C, or 94 K, which means water is useless for making rain and rivers. Instead, Titan uses a substance that exists naturally on Earth only in vapor form—methane, what we call natural gas—as the liquid phase of its hydrologic cycle. On Titan, methane clouds precipitate methane rain, which flows in methane rivers down to methane lakes (fig. 6.1). The rivers are vigorous enough to carve channels like the valley networks of Earth and Mars (fig. 6.2). Water ice is abundant, but at 94 K water ice is like granite—a rock that cannot melt or evaporate under ambient conditions. On Titan, there is no water in liquid or vapor form. The main constituent of Titan's atmosphere is nitrogen (N_2), as on Earth, and the surface pressure is 1.45 times the sea level pressure on Earth.

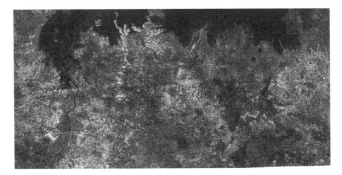

Figure 6.1. Titan lakes. Cassini sends the radar beam down at an angle with respect to the local vertical. A smooth lake acts like a mirror and reflects the beam off into space, so little or no light is received at the spacecraft. Hence rough surfaces look bright and the lakes look dark.
(Source: http://photojournal.jpl.nasa.gov/, PIA01942)

The first problem with methane on Titan is that it is slowly being destroyed at a rate that would exhaust the surface inventory in approximately forty million years.[50] This seems like a long time, but it is less than 1/100 of the age of the solar system. Destruction at this rate requires a reservoir below the surface that once had 100 times more mass than the combined mass of the lakes and the atmospheric methane vapor at present. It is unlikely that this reservoir is just about to go dry, so a reservoir of roughly this magnitude is probably still present below the surface. The depth of the lakes is unknown, although some of them are at least 8 m deep, which is the maximum depth that the Cassini radar can penetrate and still get an echo off the bottom.[51] If the atmospheric methane vapor

Figure 6.2. Titan river channels. The depth and width of the channels suggest they were carved by enormous floods. Methane is the most likely liquid. The branching suggests that rain fell over a wide area, but this is controversial. The branching is not as extensive as river channels on Earth, and could imply flow erupting from underground sources.

(Source: http://photojournal.jpl.nasa.gov/, PIA07236)

were to precipitate out, it would cover the surface with a liquid layer ~4 meters deep. A reservoir 100 times this amount is equivalent to a liquid layer 400 m deep. The nature of this reservoir and how it vents to the surface is unknown. It could be a deep methane aquifer in the soil or methane molecules trapped in an icy matrix of water molecules called clathrate hydrate. Methane clathrate hydrate forms in organic-rich sediments at the bottom of Earth's ocean, where the pressure is high and the temperature is close to 0 °C. The reservoir could take many forms, but it has to vent to the surface in order to keep supplying methane to the atmosphere. Scientists have searched images of the surface taken by the Cassini radar and find only weak evidence for volcanic landforms or vents of any sort.

Methane is being destroyed in the upper atmosphere by ultraviolet light from the Sun[52] and by energetic particles from Saturn's magnetosphere, which is the belt of electrically charged particles trapped by Saturn's magnetic field. When the energy source is ultraviolet light, the process is called photodissociation—the same process we discussed in connection with water on Venus and Mars. The UV photons have enough energy to knock hydrogen atoms off the methane molecule. As with the other terrestrial planets, the hydrogen atoms either recombine back into the parent molecule, which is methane in the case of Titan, or they escape from the gravitational field of Titan. Escape is more likely on a planet with a weak gravitational field, and Titan's surface gravity is only 14% that of the Earth. What's left behind is a fragment of a methane molecule—a carbon atom without its complement of four hydrogen atoms. These fragments are called free radicals, and they are highly reactive. They combine with each other to form more stable molecules with multiple carbon atoms—higher hydrocarbons—shedding more hydrogen atoms as they do so. Two simple hydrocarbons are ethane (C_2H_6) and acetylene (C_2H_2). Further loss of hydrogen atoms drives this process toward even higher hydrocarbon molecules, like propane (C_3H_8), methylacetylene (C_3H_4), and even benzene (C_6H_6). The fragments sometime combine with nitrogen to form nitrile compounds like hydrogen cyanide (HCN) and cyanoacetylene (C_3HN). All of these molecules have been detected in Titan's atmosphere. It is a rich organic chemistry laboratory.

The second problem with methane is finding what got left behind when the hydrogen escaped. With a steady source of methane and an ultraviolet or charged particle energy source, the escape of hydrogen irreversibly drives the chemistry toward multicarbon molecules. These molecules are not as volatile as methane—they condense to form the organic haze that blankets Titan's atmosphere. The haze particles should eventually settle out on the surface, and that presents another problem. If the equivalent of 400 meters of liquid methane has been destroyed over the age of the solar system, then roughly the same amount of heavier hydrocarbons should have accumulated on the surface. Before the Cassini spacecraft was launched, scientists were predicting it would find up to 1 kilometer of liquid hydrocarbons,[50] mostly ethane, on the surface of Titan. But the ethane ocean wasn't there. The predictions were based on the photodissociation rate, which depends mainly on the supply of ultraviolet photons and is fairly well known. It also depends on the photochemistry of the free radicals, how fast they combine to form higher hydrocarbons as opposed to recombining with hydrogen. Cassini has refined these predictions and has measured the hydrogen density in the upper atmosphere, from which it is possible to estimate the hydrogen escape rate directly. Assuming the hydrogen is coming from methane, one can estimate the rate at which the higher hydrocarbons should be accumulating at the surface. The estimated deposit over the age of the solar system is now in the 100s of m range, but still there is a discrepancy.

One possibility is that the hydrocarbon layer has percolated into the soil, especially if the hydrocarbons are liquid. If ethane is the dominant hydrocarbon, it could remain liquid at 94 K. Higher hydrocarbons—those with three or more carbon atoms, are more likely to be solid. Another possibility is that the particles are the soil. It is unlikely that the soil is weathered silicate rock like the soil on Earth. From its bulk density, Titan appears to be roughly half water and half rock. The bulk of the water may be liquid or solid, although at the surface, where the temperature is 94 K, it must be solid. Further, there is evidence that the rock has settled out to the core, leaving the water to form either a liquid ocean with an icy crust or a solid ice mantle. Thus the surface is more likely to be ice than rock, and the soil is more likely to be ice grains than weathered silicates. A mixture of solid hydrocarbons and ice is also a possibility. We do know that the soil blows around on the surface of Titan, forming into sand dunes hundreds of meters high (fig. 6.3). The Cassini radar can see the dunes, but it cannot measure their composition. The dunes do not exhibit a strong water ice signature at infrared wavelengths, leaving open the possibility that they are solid hydrocarbon particles left over from the dissociation of methane.

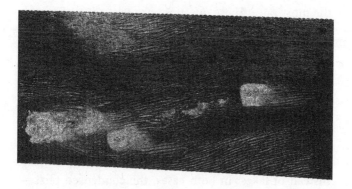

Figure 6.3. Titan dunes. The shapes of the dunes and their flow around the islands suggest a prevailing wind from the west. (Source: http://photojournal.jpl.nasa.gov/, PIA08738)

6.2 HYDROLOGIC CYCLE

Ignoring the long-term problems with methane, what is the weather and climate like on Titan today? An Earth-like planet with limited amounts of water is a possible analogue. On that planet, the water would collect at the deepest points of the ocean basins, which would be isolated from one another by the local topography. Water that fell in one basin could not flow to another basin. Rivers would deliver water to the lake at the bottom of each basin, but if the inflow were less than the evaporation rate, the lake would dry up. The questions are: If the climate stabilized with a small amount of water in the system, would the lakes be at the equator or at the poles? Where would the clouds form? Which areas would get the most rain? These questions are relevant to Titan because there is apparently only a limited amount of methane on the surface. Even though Titan has a topography of less than one kilometer, there is not enough liquid methane to cover the topography with an interconnected global ocean. A Titan fish could not swim from pole to pole, as a fish on Earth could.

Having the liquid methane immobilized within each basin makes Titan more like Mars, which has polar caps instead of lakes. In both cases, the condensed phase—water or methane—can't move, so it might as well be frozen in place. The condensed phase can evaporate, although on Mars we call it sublimation because the condensed phase is ice. On Mars, the permanent water ice has migrated to the poles because that is where the frost is coldest. Frost does form at lower latitudes, especially at night and especially during the winter, but it warms up when the Sun shines. This raises the saturation vapor pressure of the frost and causes it to sublime. Lower sublimation rate is the main reason the frost accumulates in the polar Martian regions. We say that water ice is cold-trapped at the Martian poles.

The lakes on Titan are also in the polar regions (fig. 6.1). If the lakes are in steady state and not connected by underground aquifers, the annual average of precipitation minus evaporation (P − E) for each lake basin must be zero. Rainfall can collect in rivers that flow to the lakes, but (P − E), for the basin as a whole must be zero unless there are underground aquifers. This is different from the Earth, because the oceans connect all latitudes to one another. The subtropics, which are the latitude bands on either side of the equator, are the driest, but the tropics, which are the location of the rising branch of the Hadley cell, are wet. The annual average (P − E) is positive in the tropics, negative in the subtropics, and positive at midlatitudes and the polar regions. The deficits and excesses are balanced by return flow in the oceans. On Mars, there is no return flow, since the polar caps are ice, so the annual average of (P − E) is zero at each location. Since the Titan lakes are at the poles, we have to assume that a lake at any other location would have (P − E) less than zero. Evaporation would exceed precipitation, and after a few years the lake would dry up.

Large P and small E at the poles, compared with P and E at other latitudes, favors lakes at the poles. Evaporation is smallest at the poles if the poles are the coldest places, because E is proportional to the SVP. In addition, there is evidence that precipitation is largest at the poles. During southern summer on Titan, numerous convective clouds[53] were observed over the southern pole (fig. 6.4).

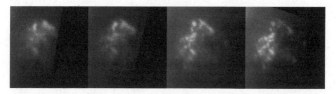

Figure 6.4. Clouds near Titan's south pole in summer. The four images span a five-hour period. Each image covers a square approximately 1600 km on a side.

(Source: http://photojournal.jpl.nasa.gov/, PIA06110)

A convective cloud is a puffy cumulus cloud, in contrast to a flat stratus cloud, and is usually a sign of vertical motion and increased precipitation. The convective activity died down after equinox, as the Sun moved into the northern hemisphere. General circulation models (GCMs) predict that the activity will start up again in the northern polar region during its summer. Since Titan's spin axis remains parallel to Saturn's spin axis as they travel around the Sun, the seasons on Titan are the same as those on Saturn, which has a 30-year orbit period. Thus each season on Titan is about 7.5 years long.

Maximum precipitation over the summer pole is explained by having the Hadley cell extend to high latitudes. On Earth the maximum precipitation is in the Inter-Tropical Convergence Zone (ITCZ), which is the rising branch of the Hadley cell. The ITCZ is where $(P - E)$ is greatest. It shifts back and forth across the equator with the seasons, always staying in the summer hemisphere although it never leaves the subtropics. This confinement occurs because Earth is a fast rotator. The Coriolis force associated with the midlatitude jet streams pushes back on the pressure force that is trying to extend the Hadley cell to higher latitudes. In contrast, Titan is a slow rotator. The rotation period is 16 days, which is also the period of the orbit. Slow rotation means that the Coriolis force is weak, so the Hadley cell extends across the equator from pole to pole. The zone of maximum precipitation is still in the summer hemisphere, as on Earth, but on Titan this maximum precipitation zone is at the pole. The general circulation models account for this effect,[54] and it seems to be borne out by cloud observations.

6.3 ENERGETIC WEATHER
IN A LOW-ENERGY ENVIRONMENT

Images sent back from the surface of Titan showed rounded boulders typical of highly turbulent streams. Moreover the valley networks (fig. 6.2) require periods of rapid discharge—floods—to carve the channels as deep as they are.[55] This is somewhat paradoxical, since the solar energy flux (power per unit area) that drives the convection is 1% that at the Earth. At the surface the percentage is even less, since the thick haze in Titan's atmosphere prevents much of the incident sunlight from reaching the ground. With less energy than on Earth to drive convection, there are two possibilities. At one extreme the convection events could be just as intense as they are on Earth and happen much less frequently. At the other extreme they could be weaker and happen about as often as convective events on Earth. The rounded boulders and depth of the stream channels suggest that the first possibility is closer to reality. It may not rain very often, but when it does, it rains very hard. Intense rainfall is consistent with the fact that Titan's atmosphere is loaded with methane—an amount equivalent to ~4 meters of liquid—compared to a few tens of centimeters of water in saturated tropical air on Earth.

If the same fraction of condensable vapor came down as rain in a single storm, the convective events on Titan would be much more intense than those on Earth.

Like Venus, Titan is a slow rotator, and its upper atmosphere exhibits superrotation. The European probe Huygens, which was carried to Saturn by the U.S. spacecraft Cassini, parachuted into Titan's atmosphere and broadcast radio signals that were received by the Cassini spacecraft. From the Doppler shift of the signals, scientists were able to determine the wind velocity at altitudes up to 130 kilometers, where the pressure is ~5 mbar. There the wind speed is 100 m s^{-1}, which is about the same as the wind speed on Venus but at a slightly greater altitude (fig. 6.5).[56,57] Earth-based spectroscopic measurements give the wind speed at ~200 km altitude where the pressure is 0.5 mbar.[58] The value is 200 m s^{-1}, which is five times the speed of the Earth's jet streams and twice the speed of the Venus winds. These high speeds are remarkable for a planet that receives only 1% of the amount of sunlight (power per unit area) as the Earth. The winds blow in the same direction that the planet rotates but much faster, so we describe the wind pattern as superrotation. Some type of wave must be bringing momentum into the upper atmosphere to maintain the superrotation against frictional drag with the surface. Several theories have been studied with general circulation models, but the observations have not established which theory is correct.

With its thick atmosphere, one might ask why Titan doesn't exhibit a stronger greenhouse effect—why its temperature structure isn't more like that of Venus (fig.

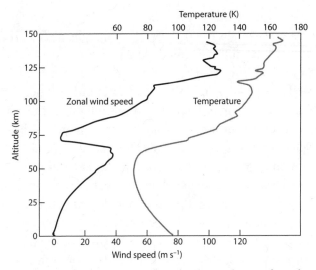

Figure 6.5. Titan temperature and winds. The winds come from the Doppler shift of the radio signal on the Huygens probe. The temperature profile comes mostly from Cassini infrared observations.

(Adapted from fig. 2 of Bird et al., 2005,[56] and fig. 1 of Flasar et al., 2005.[57] Additional source:http://atmos.nmsu.edu/data_and _services/atmospheres_data/Huygens/DWE.html)

6.5). The surface pressure, which is the weight of the atmosphere per unit surface area, is 1.45 bars. But weight is proportional to gravity, which is only 14% of the Earth's gravity, so the mass per unit area of Titan's atmosphere is 7.3 times that of Earth. This is a hefty atmosphere, and one might expect it would have a hefty greenhouse effect, but it doesn't. The 94 K temperature at the ground is not much different from that of an airless body at the orbit of Saturn. The explanation seems to be that very little of the incident sunlight reaches the ground. The organic haze absorbs most of the sunlight high in the atmosphere, and the temperature at the 200 km level is 170–180 K. The greenhouse effect requires that sunlight reach the ground and infrared radiation not get out. On Titan, only a small fraction of the sunlight reaches the ground, and the warming of the atmosphere has been called the anti-greenhouse effect.[59]

A final puzzle is that most of the lakes are in the north and only one lake is in the south. The northern lakes are much larger than the southern lake. This is not likely to be a seasonal effect, because there is not enough solar energy to evaporate the entire mass of the northern lakes and transfer that mass to the south in half a Saturn year. More likely, the asymmetry is a long-term effect that depends on the eccentricity of the orbit.[54,60] Northern summer occurs at aphelion—when Saturn and Titan are farthest from the Sun. Southern summer occurs at perihelion—when Saturn and Titan are closest to the Sun. This means that northern summers are longer and the precipitation season lasts for a longer period of time than in the south. Also, the northern summers are cooler and less intense, so the evaporation is less. This argument explains why the lakes are mostly in the north, but it hinges on the length of the summer season being the important variable. If the Hadley circulation were much more intense during the southern summer, the extra precipitation might offset the reduced length of the season. Further observations and modeling should explain this hemispheric asymmetry.

6.4 RETAINING AN ATMOSPHERE

Titan may be just above the limiting size of objects that can have a substantial atmosphere. Small gravity and proximity to the Sun make it more difficult to retain an atmosphere. The latter means the temperatures are higher and the molecules have more energy to escape. The former means less energy is required for escape. In addition, proximity to the Sun means more molecules are lost to the solar wind. If the object has a magnetic field, it may protect its atmosphere from the solar wind. And the composition of the atmosphere affects the temperature. For moons, it matters whether the parent planet had a violent phase during its formation or not. There are other factors as well. In any case, it is interesting to examine the small objects in the solar system to see which ones have atmospheres and which ones do not.

Table 6.1 lists the objects ranked according to the escape parameter λ_{esc}, which is the ratio of the gravitational binding energy $E_{grav} = GMm/R$ to the thermal energy $E_{th} = k_B T$ (box 2.2). The expression for λ_{esc} is $GMm/(Rk_B T)$. Here G is the gravitation constant, M is the mass of the planet, m is the mass of the molecule, R is the radius of the planet, k_B is Boltzmann's constant, and T is the temperature. For m, we choose atomic hydrogen because that is the atom most likely to escape—the one with the lowest binding energy. The probability that an atom has enough energy to escape is proportional to $exp(-\lambda_{esc})$. For temperature we use T in the equation $\sigma T^4 = F_0/r^2$. This is the equilibrium temperature of a black

Table 6.1.

Atmospheric escape as a means of classifying small bodies in the solar system. Bodies are listed in decreasing order of E_{grav}/E_{th}, which is the ratio of gravitational binding energy to thermal energy. A low value of the ratio means that the body is less able to retain an atmosphere over its lifetime. The ratio increases with the body's mass M, decreases with its radius R, and increases with the square root of its orbital radius r. A double asterisk (**) means the body has a thick atmosphere ($P > 6$ mbar), and a single asterisk (*) means it has a thin atmosphere ($0 < P < 0.03$ mbar). The break point for retaining a thick atmosphere is between Titan and Ganymede. Adapted from tables 1.2, 1.5, and 9.2 of de Pater and Lissauer (2010)[1]

	M (10^{23} kg)	r (AU)	R (10^3 km)	E_{grav}/E_{th}
Mars	6.418	1.524	3.390	4.76 **
Titan	1.345	9.543	2.575	3.29 **
Ganymede	1.482	5.203	2.634	2.62
Io	0.893	5.203	1.821	2.28 *
Callisto	1.076	5.203	2.403	2.08
Triton	0.215	30.069	1.352	1.78 *
Mercury	3.302	0.387	2.440	1.72
Pluto	0.132	39.48	1.137	1.49 *

surface at noon on the equator, F_o is the power per unit area of sunlight at Earth orbit, and r is the orbital radius in astronomical units (AU). In computing the temperature this way, we are neglecting the composition of the atmosphere, the rotation of the object, the albedo of the object, and many other things. For Earth, the equation gives $T = 394$ K, which is less than the temperature of the upper atmosphere where atoms are escaping. Given these limitations, it is remarkable how well the theory does.

A high value of λ_{esc} means that the probability of escape is small (box 2.2). Of the objects listed in table 6.1, Mars has the highest value of λ_{esc}, and Titan has the next highest value. This is consistent with the fact that Mars and Titan are the only objects with thick atmospheres—pressures greater than 600 Pa. Ganymede has a slightly lower value of λ_{esc} and it does not have an atmosphere. It has a higher mass than Titan and therefore has a higher E_{grav}, but it is closer to the Sun and therefore has a higher value of E_{th}. The molecules are moving faster on Ganymede and are therefore more likely to escape. Ganymede, Io, and Callisto are all satellites of Jupiter. Their order on the list depends on mass and radius, and none of them have thick atmospheres. Triton, a satellite of Neptune, is lower on the list than the Jovian moons because it is less massive. Mercury is more massive than any of the moons, including Titan, which means it has a higher binding energy. But Mercury is closer to the Sun, which means it has a higher thermal energy. The latter is more important in this case, and Mercury has a lower value of E_{grav}/E_{th}. Like Triton, Pluto has a thin atmosphere, but it ranks lower on the list because it is less massive. Earth's moon is below the bottom of the list with $\lambda_{esc} = 0.86$. The ranking in table 6.1, which is based on the single number λ_{esc}, does not always predict which objects should have thin atmospheres (pressures up to 30 Pa) and which should not. But it does show a definite break between thick atmospheres and everything else. That break is between Titan and Ganymede. Being just above the break is consistent with the evolved state of

Titan's atmosphere. The evolution, driven by escape, is a sign that Titan has struggled to retain an atmosphere during its lifetime.

6.5 THIN ATMOSPHERES

Thin atmospheres are a diverse lot. Some owe their existence to external sources like solar photons, the solar wind, and micrometeorites, which knock molecules off the surface. Others owe their existence to internal sources like volcanoes and venting. Still others owe their existence to frosts on the surface that are in vapor equilibrium with the atmosphere. Thin atmospheres tend to be ephemeral—they vary in time and space by orders of magnitude. Pressures are 30 microbars or less. Some of the thin atmospheres collapse twice a year as the polar caps wax and wane.

Triton[61] and Pluto have mostly nitrogen (N_2) atmospheres, with various amounts of methane (CH_4), carbon monoxide (CO), and carbon dioxide (CO_2). The spectrum of sunlight reflected off the planets' surfaces reveals these substances in solid form. Infrared observations show that the temperatures are 35–40 K. For the observed range of temperatures, the saturation vapor pressure of nitrogen is in the range 10–30 microbar, or 10–30 millionths of the Earth's sea level pressure. Water and carbon dioxide are detected in solid form, but the SVPs of these substances are extremely low at these temperatures, and the vapors are undetectable.

Triton and Pluto are like Mars in that their atmospheres are in equilibrium with a frost on the surface.

On Mars the frost is CO_2, and on Triton and Pluto it is N_2, but this is not fundamental to their behavior. More important is the ability of the atmosphere to buffer the swings of atmospheric pressure over the seasonal cycle. As we have seen (fig. 5.1), condensation of frost at the north and south polar caps of Mars during their respective winter seasons draws down the atmospheric pressure by as much as 33% relative to the maximum pressure. If the maximum pressure were less, the percentage drop in pressure would be greater, and it is possible that almost the entire atmosphere could disappear during the winter seasons. With maximum pressures less than 1% of the maximum pressure on Mars (10–30 μbar versus 6 mbar), and with much longer winters, it is likely that the atmospheres of Triton and Pluto collapse twice per year.[62] We haven't observed this collapse because we haven't observed the atmospheres over a seasonal cycle. Neptune's and Triton's year is 165 Earth years, and Pluto's year is 248 Earth years. The Voyager spacecraft first measured Triton's atmosphere in 1989, and the New Horizons spacecraft will fly past Pluto in July of 2015. The bending of light from stars as they pass behind the objects allows us to measure the atmospheric pressure every few years, but composition and winds are more difficult.

The only information about winds on Triton comes from the Voyager flyby in 1989. Images of the surface revealed dark streaks trending diagonally away from the summer pole with a component in the retrograde direction—opposite to the direction of rotation (fig. 6.6). On a rotating planet, such a pattern is consistent with

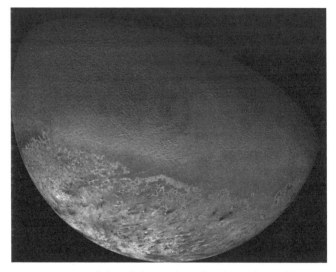

Figure 6.6. Triton disk with diagonal streaks. Plumes form when sunlight is absorbed within the ice, vaporizing the solid nitrogen and causing the gas to erupt. The prevailing wind is toward the equator, but the wind lags behind the solid surface, which rotates from right to left in this image. Hence the streaks are aligned diagonally from northeast to southwest. The pole of rotation is at bottom left.
(Source: http://photojournal.jpl.nasa.gov/, PIA00317)

flow away from the pole.[63] As the rings of air flow toward the equator, they tend to conserve angular momentum. Since the planetary object is wider at the equator than at the pole, the moment of inertia is increasing, which means that the angular velocity is decreasing. Therefore the rings of air rotate more slowly than the solid surface as they move equatorward, and they acquire a retrograde component of flow. This equatorward-retrograde flow probably occurred during the spring, when the nitrogen polar cap was evaporating and the atmosphere was flowing away from the pole.

Voyager also caught two dark plumes rising 10 km above the surface and then being blown downstream in the prograde direction.[64] The plumes were also in the summer hemisphere, at somewhat higher latitude than the streaks. The direction of the plumes suggests that the wind reversed direction from spring to summer, but further inferences are difficult. The geyserlike plumes were a surprise on an object as cold as Triton. They cannot be powered by water like geysers on Earth, but they could be powered by nitrogen or another equally volatile gas. If seasonal frost of N_2 formed a transparent or translucent layer, allowing sunlight to warm the darker material beneath it, then pressure might build up and gas might escape explosively through vents to the surface. Spiderlike features in the springtime frost caps of Mars have been interpreted in this way. The gas picks up dust and other dark material before exiting the vent and spreading out in a multi-armed pattern. On Mars the gas is CO_2, and on Triton it is N_2, but otherwise the process is the same.

On Io[65] the gas is sulfur dioxide (SO_2), which is less volatile than CO_2. At the same temperature, say at 190 K, the vapor pressure of CO_2 is 100 times greater than that of SO_2. Why the atmosphere of Io, which is farther from the Sun than Mars, should have an atmosphere of SO_2, the less volatile gas, reflects the composition of the interiors of the two bodies and the chemical reactions that have occurred at their surfaces. Sulfur dioxide gas occurs in the Venus atmosphere, and sulfate minerals occur on the surface of Mars, but SO_2 would probably not exist in vapor form on Io if there were no volcanoes. The vapor pressure at the poles and on the night side is just too low to support a global atmosphere at any significant surface pressure. In equilibrium, the atmospheric pressure might be 10^{-12} of the Earth's sea level pressure, and the rest of the SO_2 would be locked up in permanent polar caps. Instead, what we see are volcanoes constantly releasing SO_2 gas, which forms a local atmosphere of 10^{-7} bars that expands horizontally over the cold surface outside the volcano (fig. 6.7). Like the gas from a rocket nozzle or the solar wind, the gas is expanding into a vacuum on Io's night side. Expansion takes place supersonically, as the internal energy of the gas is turned into kinetic energy.[66] The atmosphere exists only over the volcanoes and over the warm frost deposited by the volcanoes. On the night side and at the poles, there is a partial vacuum. If one turned off the volcanoes, the frost would migrate to the poles and the atmosphere would collapse to a partial vacuum of 10^{-12} bar.

Walking around unprotected on the surface of Io you would be fried by the intense bombardment of charged particles from Jupiter's magnetosphere. But in your lead-lined space suit you would know which way the wind blows—from the east in the morning and from the west in the afternoon. The wind blows away from the subsolar point, where the frost is warmest and atmospheric pressure is highest. The wind also blows away from the active volcanoes, for the same reason. At night, away from the

Figure 6.7. Eruption of the volcano Pele on Jupiter's moon Io. The plume rises 300 kilometers above the surface in an umbrella-like shape. The fallout covers an area the size of Alaska. The volcano fills the image from upper left to lower right, and the vent is a dark spot just above center. To the left, the surface is covered by lava flows rich in sulfur.

(Source: http://photojournal.jpl.nasa.gov/, PIA00323)

active volcanoes, there would be no wind, and no atmosphere either, just a few molecules hopping around in the partial vacuum. This is the simplest picture, and perhaps we should qualify it. The SO_2 could break down into SO and O, and two O atoms could combine to make O_2. Since O_2 has a higher vapor pressure than SO_2, the supersonic flow might push the O_2 to the night side and hold it there. Then one would have a standing shock wave where the supersonic SO_2 from the day side slams into the O_2

on the night side. The shock wave would be something like the ring of turbulent water at the bottom of a sink when you first turn on the tap and the sink starts to fill up. The water from the tap spreads out rapidly along the bottom. The water around the edge of the sink is deeper. It is trying to flow into the center but the water flowing outward holds it back. A ring of turbulent water develops at the boundary.

Finally, there are objects like Ganymede, Callisto, Mercury, and Earth's Moon, whose atmospheres consist of individual molecules hopping around on the surface. The density of molecules is so low that the molecules don't collide with one another—they only collide with the surface. They follow ballistic trajectories while they are in the air, but they don't necessarily escape. The collisionless part of an atmosphere is called the exosphere. Small objects have exospheres down to the surface. How they get their exospheres depends on the object and its environment, and there are many possibilities. We will mention them only briefly. Mercury is closest to the Sun, and the solar wind contributes hydrogen and helium to Mercury's exosphere. Other constituents come from the planet. Sodium, potassium, and calcium have been detected, mainly because they have strong spectral lines that can be seen from Earth. These metals were probably sputtered off the surface, either by charged particles from the solar wind, by high energy photons, or by micrometeorites. Outgassing from the interior may contribute as well. Comets striking the surface are a source of volatiles, including water. The volatile molecules hop around, and some land in the permanently shadowed craters at the poles.[67] The Moon's atmosphere is like Mercury's. Sodium and potassium can be seen from Earth, and ices have been detected at the lunar poles.[68] Except for Io and Titan, the moons of the giant planets are mostly covered with water ice. Therefore solar photons and charged particles from the planets' magnetospheres release hydrogen and oxygen when they strike the surface. Since the gravity is low, the hydrogen escapes and the oxygen recombines to make O_2. Tenuous exospheres of O_2 have been detected on several satellites of the outer solar system.[65]

Objects with thin atmospheres have strong diurnal cycles. With clocklike regularity, their surfaces get warm during the day and cold during the night. The thin atmospheres are slaved to the diurnal cycles of their surfaces, so the weather is very predictable. In contract, objects with thick atmospheres have weak diurnal cycles. Their atmospheres store heat during the day and release it slowly at night. In this respect they are the opposite of the thin-atmosphere objects.

7 JUPITER THE GAS GIANT

7.1 SOLAR COMPOSITION

JUPITER[7] IS NOT ONLY THE LARGEST PLANET, BUT IT IS also the one whose composition most resembles that of the Sun. A fundamental assumption about the objects in the solar system is that they all formed from the same material with the same proportions of elements. During its first 10^5 years, the solar system was a melting pot in which material from stars with different chemical compositions was blended together. The high temperatures resulting from the release of enormous amounts of gravitational potential energy aided the melting process. Evidence for this view comes from comparing the elemental abundances on the Sun[69] with the so-called primitive meteorites.[1] Except for the most volatile elements, which are depleted in the meteorites, and lithium, which is depleted in the Sun, the elemental ratios are the same within the observational uncertainty. This agreement includes at least fifty elements in the periodic table. Their abundances relative to the common metals like Si, Mg, and Fe range down to 10^{-7}. Most of the material from the melting pot went into the Sun, but some was left behind in orbit. That material, the solar nebula, began to cool down as the rate of infalling material decreased.

Forming the planets and other objects in the solar system then became a matter of chemistry—the volatility, solubility, and ability to combine with other elements to form solid compounds. Everything that remained in the gas phase after the first few million years was likely to be blown away by the early solar wind. The giant planets, so the story goes, came closest to incorporating the entire mixture, volatiles and all, and so come closest to having solar composition. Jupiter comes closest to this ideal, although the departures from solar composition are significant and are not fully understood.

If you took a piece of the Sun and cooled it down to the temperature and pressure of Jupiter's atmosphere somewhere below the clouds, the electrons and ions would combine to make atoms and the atoms would combine to make molecules. Then let the metals and their oxides precipitate out and you would have a solar composition atmosphere. All the volatile elements would be present in the same abundance ratios as on the Sun, but they would be bound up in stable chemical compounds. Since hydrogen and helium make up more than 99% of the atoms on the Sun, the reactive elements would all combine with hydrogen. Oxygen would appear as H_2O, carbon as CH_4, nitrogen as NH_3, and so on. The nonreactive elements—the noble gases He, Ne, Ar, Kr, and Xe—would remain as they are. For reference, the ten most abundant elements on the Sun are H, He, O, C, Ne, N, Mg, Si, Fe, and S, in that order.

Table 7.1 gives the chemical abundances in a solar-composition atmosphere, along with the measured composition of Jupiter's atmosphere.[70] The first column is a list of elements, denoted by the letter X. The second column is a list of the corresponding compounds, denoted by the letter Y. The third column gives X/H for the Sun, the abundance of each column relative to hydrogen. The fourth column gives the abundance of each compound in a solar composition atmosphere—a cooled-down piece of the Sun that is basically a hydrogen-helium mixture with a 0.2–0.3% sprinkling of everything else. The fifth column gives the abundance of each compound in Jupiter's atmosphere as measured by the Galileo probe.

Jupiter is a giant ball of gas. Its density is 1.326 g cm^{-3}, which is a compromise between the low molecular mass of the constituents and the high degree of self-compression due to the planet's large gravity. Saturn's density is 0.687 g cm^{-3} even though it is made of almost the same elemental mixture. Its density is lower because its mass is only 30% that of Jupiter and the self-compression is lower. Detailed modeling[71] suggests that Saturn must have a slightly higher proportion of heavy elements (heavier than hydrogen and helium) to account for its density, even though the density is less than Jupiter's. Not surprisingly, both planets are called gas giants. Uranus and Neptune are called ice giants because they must have a larger proportion of compounds like water, ammonia, and methane to account for their higher densities despite their relative low masses. For reference, the mass of Uranus is 4.5% that of Jupiter's, and the mass of Neptune is 5.4% that of Jupiter's, yet their densities are 1.318 and 1.687 g cm^{-3}, respectively.

Table 7.1.

Abundances of elements in Jupiter's atmosphere compared with the abundances on the Sun. X stands for the element and Y stands for the corresponding compound. The third column gives the relative elemental abundances derived from solar spectroscopy. The fourth column gives the abundance of each compound in a cooled-down mixture of solar composition. The fifth column gives the abundance of each compound as measured by the Galileo probe. The sixth column gives the enrichment factor for each element—its abundance on Jupiter relative to its abundance on the Sun. An enrichment factor less than one means the element is depleted in Jupiter's atmosphere relative to the Sun. Most of the numbers are uncertain in the first or second significant digit. Adapted from table 2 of Atreya (2010).[70]

X	Y	$(X/H)_{Sun}$	Y_{solar}	$Y_{Jupiter}$	$(X/H)_{Jupiter}/(X/H)_{Sun}$
H	H_2	1.00	0.836	0.861	1.00
He	He	0.09705	0.162	0.136	0.81
O	H_2O	5.13×10^{-4}	8.6×10^{-4}	4.3×10^{-4}	0.48
C	CH_4	2.75×10^{-4}	4.6×10^{-4}	19.5×10^{-4}	4.3
Ne	Ne	2.10×10^{-4}	3.5×10^{-4}	2.1×10^{-5}	0.06
N	NH_3	6.76×10^{-5}	1.1×10^{-4}	5.7×10^{-4}	4.9
S	H_2S	1.55×10^{-5}	0.26×10^{-4}	0.78×10^{-4}	2.9
Ar	Ar	3.62×10^{-6}	6.1×10^{-6}	1.57×10^{-5}	2.5
Kr	Kr	2.14×10^{-9}	3.6×10^{-9}	7.4×10^{-9}	2.0
Xe	Xe	2.10×10^{-10}	3.5×10^{-10}	7.6×10^{-10}	2.1

The numbers for the Sun come mostly from spectroscopy.[69] Each element absorbs light at specific frequencies and leaves its fingerprint as a set of narrow spectral lines in the sunlight that we receive at Earth. The method is incredibly precise. It is possible to detect elements whose abundance relative to hydrogen is one part in 10^{12}. The noble gases do not absorb in the visible range—the electrons are so tightly bound to the nuclei that they only absorb in the extreme ultraviolet and X-ray region. You have to hit them with a high-energy photon to get them to react. This property of the noble gases is what makes them chemically inert and why they are usually found in the gas phase. They do not combine with other elements, and they do not condense to form solid ice except at very low temperatures. Estimating the abundance of the noble gases on the Sun requires indirect methods, like looking at light from the million-degree solar corona or looking at the elements of the solar wind. These methods are less reliable, and the estimated abundances for the noble gases are uncertain by 25%.

The numbers for Jupiter come from the Galileo probe, which made a fiery entry at 48 km s[-1] into Jupiter's atmosphere on December 7, 1995. The probe carried a mass spectrometer to measure the composition.[72] Attenuation of the probe's radio signal gave an independent estimate of the ammonia abundance.[73] The probe was designed to survive to the 10 bar level, but it returned good data down to 20 bars where the temperature was 420 K. By a clever bit of chemistry, the mass spectrometer was able to remove the reactive gases from the atmospheric sample, leaving only the noble gases.[74] This increased the precision to the point that the instrument was able to detect xenon, whose abundance relative to hydrogen is 2×10^{-10}. The main limitation of the probe was that it only sampled one place on the planet. Gases that are subject to cycles of evaporation and precipitation, like water vapor on Earth, could be incorrectly sampled because their abundances vary from place to place. On Jupiter there are three such gases—ammonia, hydrogen sulfide, and water vapor, which form the three types of clouds in Jupiter's atmosphere.

The last column of table 7.1 gives the ratio X/H on Jupiter divided by X/H on the Sun. These are the enrichment factors—the amount by which the abundance of that element is enriched relative to its abundance on the Sun. An enrichment factor less than 1.0 means that the element is depleted in Jupiter's atmosphere relative to solar abundance. Most of the enrichment factors are greater than 1.0, but some, like O and Ne, are less than 1.0. The values of the enrichment factors have not been fully explained. We will discuss some of the theories in the next section.

7.2 ORIGIN AND EVOLUTION

The traditional view of solar system formation says that the Sun and planets formed from the same material, having the same relative abundances of the chemical elements. While the Sun was still contracting, it went through a T-Tauri phase, named after young, solar-mass type stars, which have intense sunspot activity and powerful solar winds. In the inner solar system, the solar wind removed the gases of the solar nebula. The solids remained because they had clumped together to form solid objects large enough to withstand the solar wind. Observations of young stars indicate that the solar nebula lasts for only a few million years.[75] That is the length of time that a planet like Jupiter has to accrete nebular material. After that, only a cloud of solid planetesimals remains. With the gases gone, a planet with solar composition could not form. The traditional view seems to explain the main differences between the terrestrial planets and the giant planets. Close to the early Sun it was too hot, and the solar wind was too strong for anything but the rocky terrestrial planets to form. Further out, the gaseous components of the solar nebula lingered long enough for the giant planets to pull them in. The traditional view allows for enrichment of elements that form solid particles, either as ices or in combination with other elements, because they would remain when the hydrogen and helium were blown away. It is supported by the composition of so-called primitive meteorites, which have the same distribution of nonvolatile elements as the Sun.

Table 7.1 shows that there are problems with the traditional view. The enrichment factors for C, N, and S are between 2.5 and 5, which might have come about if 60–80%, respectively, of the hydrogen and helium were lost during the formation of Jupiter. However, the enrichment factor for O (oxygen) is less than 1, and the traditional view says that all the ice-forming elements should have similar enrichment factors. We deal with this problem below, but there are other problems. The noble gases Ar, Kr, and Xe do not form solids at the orbit of Jupiter, so they should have been carried away with the hydrogen and helium. Their enrichment factors should be 1, even if 60–80% of the hydrogen was lost, because the noble gases should have been lost along with the hydrogen.

Instead, these noble gases have enrichment factors that are similar to those of C, N, and S. These enrichments are a puzzle. The explanations that have been proposed involve bringing these gases in later,[76] either as noble gas ices from the outer edge of the solar system where temperatures are 30 K or below, or else riding on water-rich comets in which noble gases were trapped in clathrate hydrate cages. Either way, this component of Jupiter's atmosphere is secondary—it was not there when the planet formed.

The above explanation may work for Jupiter, but it causes problems for terrestrial planets, because some of the icy comets would have found their way to the inner solar system. If the total icy comet mass was large enough to alter Jupiter's atmosphere, it would have been large enough to alter the terrestrial planet atmospheres. The problem is that the terrestrial planets are severely depleted in the noble gases relative to C, N, S, and O. Their argon is ^{40}Ar, which comes from potassium decay and is different from the primordial ^{36}Ar found on the Sun and Jupiter. Thus the icy comets proposed to explain the enrichments of noble gases on Jupiter are inconsistent with the depletion of noble gases on the terrestrial planets. The atmospheric abundances of the terrestrial planets reflect a complex history of outgassing from the planetary interiors, delivery of volatiles by comets and meteorites, reactions with crustal rocks, and escape from the top of the atmospheres. They are secondary atmospheres. We know there is very little volatile material left over from the solar nebula, because the noble gas abundances are so low. Thus it is puzzling that the noble gases are enriched relative to solar composition in Jupiter's atmosphere. The Galileo probe results have complicated our picture for the origin of atmospheres throughout the solar system.

Neon apparently has a different history from that of the other noble gases, since it is depleted with respect to solar composition by almost a factor of twenty. The explanation could be that neon has settled into the core. According to this theory, the planet as a whole is not depleted in neon. The modest depletion of helium (enrichment factor = 0.81) is part of the proposed explanation. Theoretical work on the equation of state of hydrogen-helium mixtures indicates that the hydrogen becomes metallic at a pressure of several megabars and that the helium becomes immiscible in the metallic hydrogen. In other words, helium might condense to form raindrops, which could settle toward the core like raindrops in Earth's atmosphere.[77] Neon dissolves in the helium raindrops, and is carried down with them. The raindrops form at a level that is 20% of the way down to the center of the planet (at 80% of the radius). Since the fluid is mixed by convection currents above this level, the depletion extends up into the atmosphere where it was measured by the Galileo probe. Neon is much less abundant than helium, so it is more affected by this process. Therefore its degree of depletion is much greater than helium's.

According to the models, the process is still going on. As the planet cools, the SVP of the helium raindrops decreases, and this causes more helium to precipitate. The

He/H ratio in Jupiter's atmosphere decreases as a result. Indirect support for this theory comes from Saturn, which seems to have a smaller He/H ratio in its atmosphere than Jupiter. Saturn is a less massive planet and therefore it cools faster. Precipitation of helium is further along on Saturn than it is on Jupiter, and the neon depletion should be almost complete. There has never been an instrumented probe into Saturn's atmosphere, so the Saturn evidence is subject to large uncertainty.

7.3 CLOUDS AND VERTICAL STRUCTURE

Water is the remaining problem shown in table 7.1. If C, N, and S have enrichment factors between 2.5 and 5.0, one might expect the same ratio for O. However, the depletion of O (enrichment factor less than 1.0) may have a meteorological explanation—the Galileo probe seems to have parachuted into the Sahara Desert of Jupiter. Jupiter's atmosphere is normally covered with clouds, but the probe went into a cloud-free area—a so-called 5-micron hot spot (fig. 7.1).[78]

The atmospheric gases do not absorb radiation at 5 microns, so radiation from warmer, deeper levels can escape to space if there is a hole in the clouds. Hence they appear hot. Absence of clouds means low abundance of condensable gases. In other words, the hot spots are dry, which might explain the low water-vapor abundance and therefore the low O/H ratio at the Galileo probe site.

Figure 7.2 shows the vertical structure of an idealized Jupiter atmosphere.[79,80] The temperature profile is from

Figure 7.1. Jupiter disk showing thermal radiation emerging from holes in the clouds at 5 microns wavelength. The Galileo probe entered the atmosphere in a hot spot like the one in the center of the disk.

(From plate 1 of Ortiz et al. [1998])[78]

sensors on the probe, but the clouds are obtained from a model assuming enrichment factors of 3.0 for the three condensable gases NH_3, H_2S, and H_2O. The abundances are three times the values given in the fourth column of table 7.1, and they apply to the atmosphere below the clouds. Using the temperature profile and the assumed threefold enrichment factors, one can compute the cloud base for each of the three condensing gases. Ammonia is the most volatile gas. Its vapor pressure is high, and it

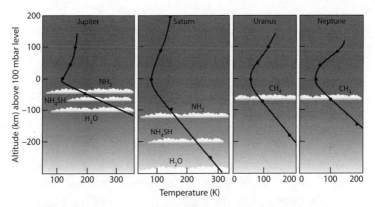

Figure 7.2. Vertical structure for the four giant planets. The zero of altitude is where pressure equals 100 mbar. Black dots are where the pressure is 1 mbar, 10 mbar, 100 mbar, 1 bar, and 10 bar. The temperature profile is the smooth curve and uses an adiabatic extrapolation below the ~1 bar level. Cloud base assumes enrichment factors of 3, 5, 30, and 30 relative to solar composition for Jupiter, Saturn, Uranus, and Neptune, respectively.

(Adapted from fig. 8 of Ingersoll, 1999)[80]

requires the lowest temperature to condense. Thus the ammonia cloud base is above the other two. Hydrogen sulfide is thought to combine with ammonia at intermediate levels to form cloud particles of NH_4SH. Water is the least volatile gas, and it condenses below the other two gases. Cloud base depends only on the temperature profile and the assumed enrichment factors. Cloud top depends on factors like the speed of the updrafts, and is highly uncertain. Rapid convective updrafts could carry the water cloud particles up to the top of the ammonia cloud, or the cloud particles could form a thin stratus layer near cloud base. Very dry air, like the 5-micron hot spot, could have no clouds at all.

Before the Galileo probe went in, we had a naïve view of what the probe would find—constant abundance ratios of the gases up to the base of the clouds followed by rapid decrease of each condensable gas above its cloud base. For 3 times solar enrichment factors, the base of the water cloud is at 7.5 bars, that of the NH_4SH cloud is at 2.5 bars, and that of the ammonia cloud is at 0.7 bars. Below these levels we imagined that the probe would be sampling the bulk composition of the planet. We expected enrichment factors greater than 1.0 at the base of the clouds. Instead the probe found enrichment factors of ~0.1 where we expected the base of the clouds. The abundance ratios slowly increased as the probe descended. When the probe reached 7 bars, the ammonia abundance had finally leveled off at 4.9 times solar. When the probe reached 16 bars, the H_2S abundance had leveled off at 2.9 times solar. When the probe stopped transmitting at 20 bars, the water abundance was 0.48 times solar but it was still increasing.[72]

The delayed increase of the abundances as the probe descended was unexpected. One explanation is that the 5-micron hot spots are giant downdrafts where the condensable gases have been pushed down to where the pressure is ~5 times its normal value. Thus it is possible that the water vapor abundance would have leveled off at 3 times solar when the pressure reached ~40 bars. The probe may have entered the Sahara Desert of Jupiter, and the naïve view may prevail over the rest of the planet, with 3 times solar abundance ratios up to cloud base. Determining the water abundance is important, first because icy comets are likely to have played a role in delivering volatiles throughout the solar system, and second because water is important for the weather and climate of Jupiter. Oxygen is the third most abundant element on the Sun, after hydrogen and helium, so water should be the most abundant compound after H_2.

In some ways, the vertical temperature structure of Jupiter's atmosphere resembles that of Venus—an adiabatic lapse rate at depth, clouds at intermediate depths, and a subadiabatic atmosphere above the clouds, where the infrared photons escape to space. The adiabatic lapse rate implies convection currents that are carrying heat up from below. On Venus, the source of this heat is sunlight reaching the ground—the remaining part of the incident sunlight that is not reflected back to space or absorbed high in the atmosphere. On Jupiter, the source is heat stored in the interior of the planet from the time of solar system formation. As a planet accreted smaller objects, its gravity accelerated the objects to higher and higher speeds. The resulting collisions turned kinetic energy into heat, which got buried when more objects came in. Later, convection currents carried the heat to the surface, where it is radiated away, but this can be a slow process. Terrestrial planets have warm surfaces and large surface-to-mass ratios, so they lose their internal heat quickly. Giant planets are the opposite, so they lose heat slowly. Jupiter currently radiates 1.7 times more energy than it absorbs from the Sun. Therefore a fraction

0.7/1.7 of the emitted radiation comes from internal heat. A billion years ago, the fraction was larger, but the planet has cooled during the intervening years. For Saturn, Uranus, and Neptune, the ratios of emitted infrared radiation to absorbed sunlight are 1.9, 1.1, and 2.8, respectively. Uranus is a bit of a mystery because its internal emitted power is so low. For reference, the Earth's internal heat flow is $\sim 3 \times 10^{-4}$ times the value needed to balance the absorbed sunlight.

7.4 HORIZONTAL STRUCTURE AND LIGHTNING

As on Venus, the emitted infrared radiation is largely independent of latitude[81] (fig. 7.3). This contrasts with Earth, where the emitted radiation decreases significantly from the equator to the poles, in response to the extra sunlight absorbed at the equator. We explained the difference between Venus and Earth, first as due to the greater mass of the Venus atmosphere, and, second as due to the slower rotation of the Venus atmosphere. Jupiter is a rapidly rotating planet, so the second explanation doesn't apply. Jupiter is also a fluid planet, and the fluid has enormous heat capacity and is able to transport heat readily between equator and poles. Whether the heat transport is taking place in the atmosphere or in the interior is an open question. Convection of internal heat is sensitive to small temperature differences. When the lapse rate is slightly subadiabatic, the heat flux is zero, and when the lapse rate is slightly superadiabatic the heat flux is large. This means that a slight temperature rise in

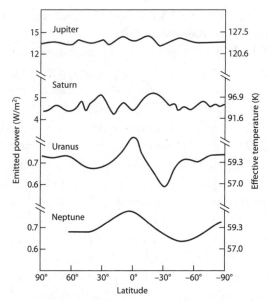

Figure 7.3. Radiative energy fluxes vs. latitude for the four giant planets. The fluxes are given on the left scale, and the equivalent black body temperatures are given on the right. The small-scale variations are associated with the zonal jets. The large-scale variations, i.e., the equator-to-pole differences, are small. In this respect the giant planets resemble Venus more than they resemble Earth (fig. 3.4).

(Adapted from fig. 5 of Ingersoll, 1990)[81]

the equatorial atmosphere will shut off the internal heat, and a slight temperature drop in the polar atmosphere will increase the internal heat by a large amount. Thus the absorption of sunlight at the equator causes internal heat to come out at the poles, leaving the atmosphere with an undetectably small equator-to-pole temperature gradient. The upward flux of internal heat acts as a thermostat for the atmosphere.[82]

There is a level near the base of the water cloud where temperatures are near the freezing point of water. This would be the level of rain, snow, and hail, if the clouds behaved like water clouds on Earth. It would also be the level of electrical activity, where collisions between small liquid droplets and larger ice particles carry positive charge aloft and negative change downward. Thus one might expect lightning at this level, and indeed, that is where lightning is observed at Jupiter (fig. 7.4). The lightning flashes stand out in images of the night side, because there is no background of small-scale clouds in the image. If one superposes day-side and night-side images, it becomes clear that the lightning occurs in a type of storm that resembles a terrestrial thunderstorm.[83] The storm clouds are brighter than the surrounding clouds, and they extend to high altitudes. They expand laterally when they first appear, and they last for a day or two. The Jovian thunderstorms range up to 1000 km in diameter, which makes them more like thunderstorm clusters on Earth. Also, they are more isolated. About 2000 thunderstorms exist at any one time on Earth, whereas only 50 thunderstorms exist on Jupiter at any one time, and Jupiter has over 100 times the surface area. The typical distance between storms is comparable to the distance halfway around the Earth.

The lightning flashes are impressive. Assuming the light goes out in all directions and is not attenuated on the way to the spacecraft, the total optical energy in a

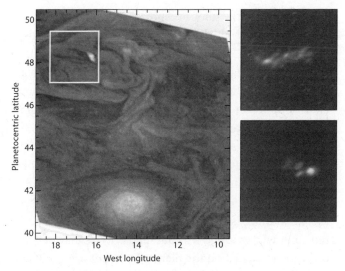

Figure 7.4. Lightning in Jupiter's atmosphere. The left panel is an image taken by the Galileo camera of clouds on the day side. The region outlined by the white square was imaged 1 hour 50 minutes later on the night side. The two boxes on the right show the region in the white square at night. The boxes on the right have been enlarged to show the multiple lightning flashes in greater detail. The top and bottom boxes have exposures of 166 s and 38 s, respectively.
(From fig. 6 of Little et al. [1999])[83]

single flash is greater than the most energetic flashes on Earth. A typical storm emits one of these large flashes every ten seconds. Small flashes are below the sensitivity of the camera, and their number is unknown. The photons emerge from the top of the ammonia cloud (fig. 7.2) as circular patches 150 km in diameter. More than 10% of the light lies outside a ring 300 km in diameter. The width of the patches is proportional to the depth of the lightning, since the photons spread out more the farther they have to go. Models of light scattering in the clouds[83] put the flashes ~100 km below the tops of the ammonia cloud, that is, near the base of the water cloud and near the level where the temperature is close to 0 °C. This provides support for the assumption that water is abundant on Jupiter, despite its poor showing (enrichment factor = 0.48) at the Galileo probe site.

Jupiter's clouds are colored. The Great Red Spot looks dark in a blue filter, that is, in a filter that transmits only blue light (fig. 7.5).[84] It has about the same brightness as the surroundings in a red filter, so the spot has a dark red color when viewed without filters. There may be only one coloring agent on the planet. Different brightness of the red material and different brightness of a white material, blended in different proportions, explains most of the contrast in cloud images. It is difficult to identify the coloring agent without doing a chemical analysis of the cloud particles, and this hasn't been done. The difficulty arises because "dark in blue, bright in red" is not very specific, and solid particles do not have narrow spectral signatures like those of gases. Further, the main clouds in figure 7.2 are white, so one needs a trace species like sulfur, phosphorous, or multicarbon organic molecules to provide the color. Such species are likely to be out of chemical equilibrium with their surroundings, and might have been created either by ultraviolet light, by lightning, or by rapid vertical transport between layers of different temperature. The color of the clouds is correlated with the dynamics of the atmosphere. Rapid

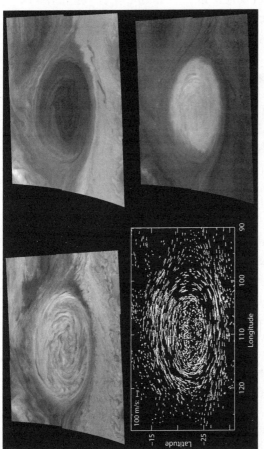

Figure 7.5. The Great Red Spot of Jupiter through filters of different wavelengths. *Clockwise from upper left:* near-IR (757 nm), violet (410 nm), methane absorption band (889 nm), and cloud-tracked winds from fig. 2 of Dowling and Ingersoll (1988).[84] The top two images show light reflected off the main cloud. The cloud particles absorb at short wavelengths, including violet, giving the Red Spot its color. Methane gas absorbs in the 889 nm filter, obscuring the main cloud and making the high clouds look brighter than the absorbing gases nearby. The winds blow in a counterclockwise direction around the Red Spot at speeds over 100 m s⁻¹, with slow-moving clouds at the center.

(Source: http://photojournal.jpl.nasa.gov/, PIA00721)

updrafts tend to be white, deep clouds tend to be dark, and spinning vortices are more or less red depending on the direction in which they are spinning. The embarrassing little secret is that we don't know precisely what the colors are made of.

7.5 HOT AIR BALLOONING

If you wanted to spend some time on Jupiter, you would probably need a hot air balloon. An airplane would get you around, but it would soon run out of gas. Actually, you wouldn't use gasoline because it wouldn't burn in the hydrogen atmosphere. Instead you would use oxygen, chlorine, or some other kind of oxidizer. The air from your breathing apparatus would burn if it came in contact with the atmosphere. In a hydrogen atmosphere, oxygen is an explosive gas, just as hydrogen is an explosive gas in Earth's atmosphere. Refueling the airplane would be a problem because there's no place to land— there is no land and no place to store the fuel. Probably you would have to import your fuel. Assuming you only had one chance to bring it in, you would probably have to store your fuel in the basket of a balloon.

Ballooning is difficult in a hydrogen-helium atmosphere, but it's not impossible. A balloon filled with pure hydrogen gas would be lighter than the hydrogen-helium mixture, and that would give you some buoyancy. Separating the hydrogen from the helium would be difficult, so you would probably have to import the hydrogen. Hydrogen is the lightest gas, so many options are ruled out. A hot air balloon that burned oxygen and released the combustion products into the bag would not work because the combustion product, water, would have a higher density than the ambient atmosphere. The molecular mass of water is 18, and the molecular mass of Jupiter's hydrogen-helium mixture is 2.3. A better design would use the ambient hydrogen-helium mixture for the gas inside. This is the "green" approach, since it uses locally available raw materials. You could keep the gas hot in several ways. One is with a heat exchanger: The heater, mounted at the top of the balloon, produces hot water vapor that flows down along the outside skin of the balloon, heating the gas inside but not mixing with it. The other way would use the ambient radiation fluxes to keep the balloon warm. This is the "double green" approach, since it uses ambient air and ambient energy sources. The air inside would have to be warmer than the ambient air, since they both have the same composition. During the day you would want the balloon to be black, to absorb the incident sunlight. During the night you would want the bottom of the balloon to be an infrared absorber and the top of the balloon to be an insulator—a material that loses heat slowly. The balloon would gain net energy by absorbing radiation from the warm layers below and by not radiating to the cold layers above.

There are a few technical details that have to be worked out, but let us ignore them and take an imaginary balloon flight through the atmosphere of Jupiter. We start out in full sunlight above the top of the ammonia cloud. Full sunlight is misleading, because Jupiter is 5.2 times farther from the Sun than Earth, and the Sun's rays are 27 times less intense. Also, we are close to the base of the stratosphere (fig. 7.2), and the temperature is incredibly cold—below 100 K, which means −173 °C or colder. The consolation is that the view is fine and we might see the towering clouds of a thunderstorm off in the distance. Actually the odds of this are rather small, since thunderstorms are a rare event and quite far apart on Jupiter.

If we are lucky we might catch one of the twenty-odd jet streams,[81] half of which flow eastward and half of which flow westward, each at its own particular latitude (fig. 7.6). The fastest jet streams flow at 120–160 m s^{-1}, which is three or four times faster than the Earth's jet streams. Catching one of the jet streams would allow us to change longitude rather quickly. So if the Great Red Spot were around on the other side of the planet, we could take an eastward-moving jet stream and approach it from the west, or we could take a westward-moving jet stream and approach it from the east. The trip halfway around the planet would take a few weeks. Of course, catching the right jet stream means moving to the right latitude, and that could be tricky. The dominant winds blow east and west, but there are turbulent eddies and waves that bring air across latitude circles. In a balloon we would have to take advantage of these features, waiting for an eddy going north or south, depending on where we wanted to go, and then getting into it. That involves changing from one air mass to another, which means either turning our balloon into a dirigible with a motor, or turning it into a kite. The latter is the "ultimate green" approach, but it

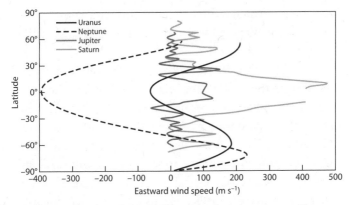

Figure 7.6. Zonal winds on the four giant planets as measured by cloud tracking. Except for Saturn, periodic variations of the internal magnetic field define the reference from which the winds are measured. For Saturn, the variations come from the magnetosphere and are not a reliable measure of the interior rotation. Despite the uncertainty in Saturn's winds, it is likely that Neptune has the fastest winds in the solar system.

(Adapted from fig. 4 of Ingersoll, 1990)[81]

needs wind shear to work. Imagine you were floating at one level at one speed, and the wind ten kilometers below was blowing at a different speed. That difference is called wind shear. You could lower your kite into the current down below and let it pull you in a different direction. By setting the kite at an angle to the wind, you could "tack" like a sailboat and go across the wind direction. It works whether the kite is above you or below you. With enough wind shear and a little skill, you could sail anywhere you wanted to go. We don't know if it would be a rough ride or a smooth one, because roughness depends on small-scale turbulence, which is difficult to measure from outside the atmosphere. It's possible that the fastest jet streams would give the smoothest rides.

So let us drop 100 km down to a more comfortable level, near the base of the water cloud where the temperature is 300 K. There's not much sunlight down there because the clouds above have either absorbed it or reflected it back to space. However there is rain, and there is an occasional thunderstorm, complete with brilliant lightning, even by terrestrial standards. We don't know if there is perpetual drizzle, because we can't see down to this level from outside the atmosphere. If there is rain, it might wet the surface of our balloon and weigh it down. That would be dangerous, since we are trying to stay afloat with minimum buoyancy in this hydrogen-helium atmosphere. The rain that doesn't land on the balloon has nothing else to land on, so it falls into the warmer layers below and eventually evaporates.

Before dropping further, we might travel around to visit the natural wonders of this world. It's a big world. Traveling from equator to pole is the same as traveling three times around the Earth. First we fly over to the Great Red Spot (GRS), and we find it is surprisingly calm inside (fig. 7.5). The winds blow around the periphery at 100 m s^{-1}, but they do not appear turbulent. We are in a southern hemisphere anticyclone as large as two Earths, so the circulation is counterclockwise around the outer perimeter. If we leave the Red Spot to visit the pole, we find that the temperature does not get colder (fig. 7.3). In fact, the pole has the same temperature as the equator[81] even though sunlight beams down directly overhead at the equator and is never more than 3° above the horizon at the pole. At some level, either in the clouds or way below them, the atmosphere must be transporting heat poleward, but it's hard to see it happening when you are limited to one balloon.

Let us go down another 100 km to the level where the Galileo probe died, to pay homage to our species' first attempt to explore Jupiter from within. The pressure at this level is 20 bars and the temperature is 420 K, which is 147 °C or 297 °F. In other words, the probe died in a high-pressure oven. It's very dark in this oven. Sunlight normally doesn't penetrate through the overlying clouds, which extend almost 200 km above our head. Even in a 5-micron hot spot, where the clouds are thin or nonexistent, the gases don't let the sunlight through to this level. As we sit in our hot air balloon, maintaining buoyancy by selectively absorbing the infrared radiation from warmer layers below, we realize how dangerous this journey is. If the balloon started to leak and lose its hot air, we would sink. And there is no safe landing on Jupiter. The atmosphere could just as well be bottomless. At some point the gaseous hydrogen-helium mixture becomes a supercritical liquid, and then it becomes a metal, where electrons wander freely between the positive nuclei. The temperature at the center of Jupiter is 30,000 K, but we would never reach that level, even if our balloon failed. Long before reaching the center, we would melt and then vaporize, our atoms becoming part of the planet. We would be just a few drops in a very big bucket.

8 JUPITER WINDS AND WEATHER

8.1 LONG-RANGE FORECASTING

A PERSONAL ANECDOTE ILLUSTRATES THE AMAZING properties of Jupiter's weather. I was a member of the imaging team that had to prepare for the encounter of Voyager 1 with Jupiter on March 5, 1979. We had to decide where to point the camera during closest approach in order to photograph the most interesting features at the highest possible resolution. At closest approach, the field of view of the camera would be much too small to photograph the whole planet, so we had to choose the interesting features and predict where they would be. The engineers wanted our predictions four weeks in advance, in order to translate our requests into language that the spacecraft could understand and not disrupt the delicate maneuvers that would slingshot the spacecraft to Saturn. In other words, we had to make a four-week weather forecast of the storms on Jupiter.

A four-week forecast of the storms on Earth is not feasible now and may be theoretically impossible. The traveling storms don't last that long, and they behave erratically during their short lifetimes. The stationary highs and lows last longer, but they are locked to the continents and oceans. Jupiter doesn't have continents and oceans—only a massive atmosphere down to its center. However, Earth-based observers had been watching the Great Red Spot ever since Robert Hooke discovered it in 1664. Telescopes had been improving when Hooke made his observations, and it is likely that the Red Spot had been in existence long before he discovered it.

Since there are no solid surfaces to organize the flow, the Red Spot is an atmospheric structure—a storm— that is free to move about under the laws of fluid dynamics. Those laws, as we know them on Earth, lead to turbulence, chaos, and limited predictability. By comparison, the Red Spot is well behaved. It stays in one latitude band, rolling like a ball bearing between two conveyor belts—a westward current to the north and an eastward current to the south. All the large-scale features are remarkably constant (fig. 8.1). Consistent with the rolling motion, small spots inside the Red Spot move around counterclockwise (fig. 7.5). The Red Spot itself usually moves westward relative to the magnetic field, which we assume is tied to the interior of the planet, but sometimes it slows down and moves eastward. These changes of velocity take place over periods of years, and do not prevent forecasts over a four-week period. The Voyager imaging team eventually used junior high school mathematics—a straight line on graph paper—to extrapolate the positions and make a beautiful mosaic of high-resolution images that covered the Red Spot from end to end and top to bottom.

The big surprise for atmospheric scientists during the Voyager encounter was not our success in predicting the

Figure 8.1. Voyager image of Jupiter in 1979 compared to a Cassini image taken in 2000. The jet streams blow east-west and are strongest at the edges of the light and dark bands. The colors of the bands are remarkably constant from 1979 to 2000. Even when the colors do change, the latitudes and speeds of the jets remain constant.

(Sources: http://photojournal.jpl.nasa.gov/, PIA00343 and http://photojournal.jpl.nasa.gov/, PIA04866)

locations of the Red Spot and other large-scale features. We had been following them for years using Earth-based telescopes, and we knew how predictable they were. For instance, the counterclockwise band just to the south of the Red Spot's band had split into three pieces over a two-year period in 1938–39. The pieces consolidated into three white ovals, each rotating counterclockwise, and the spaces between filled with featureless clouds—or so we thought. The ovals were still there in 1979, but what surprised us was the enormous activity within the featureless areas, the areas outside the Red Spot and the three white ovals. The activity had been difficult to see because it involved small-scale features that were below the resolution of Earth-based telescopes. As Voyager closed in, its camera saw clouds popping up, churning around, and mixing on time scales of hours. In our eyes, the formerly featureless areas had become turbulent convective regions. The big surprise was how the large ovals could last for decades or centuries in the midst of all this chaos.

After the Voyager encounters, many of us believed that there was some special property of Jupiter's atmosphere that allowed large vortices to exist for decades or hundreds of years. Maybe it had to do with the winds below the clouds, which we couldn't see. Or maybe large vortices required special initial conditions—unlikely events that created them. In my first attempts at modeling the Red Spot, I carefully chose its initial shape and winds to match the observations.[85] But soon it became clear that we didn't have to be so careful. Long-lived vortices were easy to make—they formed spontaneously. Different people with different assumptions were all getting long-lived vortices in their numerical models.[86,87,88] One could start off with an unstable shear flow and a small perturbation. The perturbation amplitude would grow with time, and the shear flow would spin off numerous waves and eddies. The eddies would start merging and soon there was only one eddy left, a giant oval like the Great Red Spot that lasted indefinitely. Since the ovals existed in a computer, one could experiment with them. For instance, the ovals tended to be elliptical with the long axis east-west, just like the ovals on Jupiter, but one could pick them up and turn them north-south—crosswise to the flow. The oval would shake violently and then regain its preferred east-west orientation. Or one could place two ovals at slightly different latitudes so they were moving at slightly different speeds, with the fast one catching up to the slow one. The collision usually led to a merger, although the resultant vortex had to shed some material before settling down. In other words, long-lived vortices form spontaneously in unstable shear flows, and they are stable once they have formed. The Jovian ovals may need a little energy to counteract friction, but they seem to acquire it by merging with smaller vortices.

These models were much simpler than the models used in numerical weather forecasting, which have sunlight and infrared radiation, continents and oceans, clouds, and precipitation. Maybe the Earth is too complicated to have long-lived storms—there are too many competing processes. In particular, the Earth is stormier than Jupiter. The large storms, like the winter storms at midlatitude, draw their energy from the equator-to-pole temperature gradient, which represents a source of potential energy. The winter storms are constantly forming and dying and eclipsing older storms, and the sunlight falling on the equator is constantly renewing the potential energy. This is different from the ovals in the Jupiter simulations, which have no competition once they have formed. In fact, there are analytic models from before the computer age that have stable ovals in shear flows.[89] These pencil-and-paper solutions to horizontal flow without viscosity were once viewed as a curiosity. Now they are used to describe nature on a grand scale. The message is that maybe Earth is special, and that long-lived vortices are the rule and not the exception.

8.2 ROTATING FLUIDS

Earth's rotation is what makes hurricanes spin counter-clockwise in the northern hemisphere and clockwise in the southern hemisphere. It is what makes the jet streams flow west to east and not from equator to pole or from day to night. Rotation makes the planet oblate, and it gives rise to the Coriolis force, which acts on objects that are moving relative to the rotating system. The giant planets are rapid rotators, so rotation is important in their atmospheric motions. Moreover, the colored clouds make it easy to track the motions. Because they are so photogenic, the giant planets are often regarded as laboratories for studying the dynamics of rotating fluids.[90]

Jupiter and Saturn are noticeably oblate (box 8.1). At Jupiter's equator, the distance from the center of the planet to the 1-bar pressure level is 71,492 km, but at the pole the distance is 66,854 km. The oblateness, or flattening (difference between equatorial and polar radii divided by the equatorial radius) is 6.5% for Jupiter, 9.8% for Saturn, and 0.34% for Earth. Oblateness arises from the planet's rotation and is proportional to the ratio of the centrifugal force to gravity. The centrifugal force is trying to throw things away from the axis of rotation, and gravity is trying to pull things back into a spherical shape. Since the centrifugal force (actually force per unit

---

Box 8.1.
Rotating Systems

Our bodies can sense acceleration. One way to describe the sensation is with inertial forces—the extra weight when the elevator you are in slows in its descent and the centrifugal force when your automobile goes around a turn. Inertial forces also affect objects on a rotating planet. One such force is the centrifugal force, which gives rotating planets their oblateness—the ratio $(R_e - R_p)/R_e$ where R_e is the equatorial radius and R_p is the polar radius. If there were no winds, a rotating fluid planet would assume an equilibrium shape in which the centrifugal force and gravity are exactly balanced, with the equatorial radius greater than the polar radius. Even solid planets come close to the equilibrium shape, because the rocks cannot support the imbalance. A spherical Earth spinning at its current rate would quickly collapse back into its current equilibrium shape with $R_e - R_p = 21$ km. The strength of the rocks determines the heights of the mountains, which range up to 9 km above sea level.

The other inertial force that arises in a rotating system is the Coriolis force. It only acts on objects that are moving relative to the planet. In a system rotating at a constant rate, Newton's second law $a = F/m$ can be written

$$\vec{a} + 2\vec{\Omega} \times \vec{v} + \vec{\Omega} \times (\vec{\Omega} \times \vec{R}) = \vec{F}/m + \vec{F}_N/m$$

Here $\vec{a}$ and $\vec{v}$ are the acceleration and velocity that we measure using coordinates that rotate with the planet. $\vec{R}$ is the vector from the center of the planet to the mass m. We have split the force into two parts. $\vec{F}_N$ is Newtonian gravity—the pull of all the masses in the planet, and $\vec{F}$ is all the other forces acting on the mass m. The third term on the left is the centripetal acceleration. If we put it on the right side of the equation it becomes

---

the centrifugal force per unit mass and can be combined with Newtonian gravity. The resultant is $-g\hat{k}$ where $\hat{k}$ is the vertical unit vector and g is the gravitational acceleration that we measure. By definition, the vertical unit vector is perpendicular to the equilibrium surface. The second term on the left is the Coriolis acceleration, and if we put it on the right it becomes the Coriolis force per unit mass.

mass) at the equator is $R\Omega^2$, where R is the equatorial radius and Ω is the angular velocity, the oblateness is of order $R\Omega^2/g$.

The giant planets are more oblate than Earth because they are large and they rotate faster. The angular velocity of rotation is 2π divided by the length of the day, which for Jupiter is 9 h 55 m 29 s. Since Jupiter is a fluid planet with no solid surface, the period is determined from the daily wobble of the magnetic field. The field is fixed relative to the planet's interior, but it is tilted by 10° with respect to the rotation axis, so it appears to wobble when viewed from a nonrotating reference frame. Saturn's magnetic field has no measureable tilt, so the length of day is less certain. Atmospheric periods, which are based on the cloud motions at specific latitudes, range from 10 h 11 m to 10 h 41 m. Saturn's rotation period is probably within this range. Saturn's oblateness is larger than Jupiter's mainly because its gravitational acceleration g is smaller, which follows from its smaller mass.

The centrifugal force is a static force in the rotating system and is indistinguishable from gravity for stationary objects. The surface of a lake and a weight hanging from a string feel the resultant of two forces—the centrifugal force and the gravitational attraction of the planet's mass. Mean sea level reflects this balance too. The reason we all don't slide to the equator under the action of the centrifugal force is that we already did—the entire Earth has slid. We don't worry about the centrifugal force because it is incorporated into gravity and into our definition of a level surface like mean sea level. The Coriolis force, which acts on mass that is moving relative to the rotating reference frame, upsets this balance and is therefore a major factor in the dynamics of rotating fluids (box 8.2).

Imagine a ball tossed back and forth on a merry-go-round that is turning counterclockwise like the Earth viewed from the North Pole. When the ball is in the air it travels in a straight line, but the merry-go-round rotates below it. As viewed from the rotating system, the trajectory appears to curve to the right, and we would say there was a force to the right of the velocity. This is the usual explanation for the Coriolis force, but it is not quite correct. The ball is actually feeling two forces when viewed in the rotating system. One is the Coriolis force, which acts to the right of the velocity, and the other is the centrifugal force, which acts radially outward from the axis of rotation (box 8.3).

To balance the Coriolis force, a horizontal air current blowing in a straight line in the northern hemisphere must have another force pushing to the left. In a planetary atmosphere, this other force is pressure. The high pressure is on the right of the current, and the

Box 8.2.
Equations of Motion

With the centrifugal force incorporated into gravity, the equation of motion for a rotating fluid becomes

$$\frac{d\vec{v}}{dt} + 2\vec{\Omega} \times \vec{v} = -\frac{1}{\rho}\vec{\nabla}P - g\hat{k}$$

The first term is the acceleration—the rate of change of velocity measured in the rotating system. The second term is the Coriolis acceleration. The third term is the pressure force per unit mass. The fourth term is gravity—the resultant of Newtonian gravity and the centrifugal force per unit mass. If the left side were zero, the two terms on the right would be a generalization of hydrostatic equilibrium to three dimensions. Consider a thin sheet of fluid between x and $x + dx$. If there is a gradient of pressure in direction x, then $-(\partial P/\partial x)dx$ is the net pressure force in the x direction per unit area. If the area is $dydz$, then $-(\partial P/\partial x)$ $dxdydz$ is the net force on the volume $dxdydz$. The mass inside the volume is $\rho dxdydz$, so the net force per unit mass is $-(1/\rho)$ $(\partial P/\partial x)$. Generalizing this to three dimensions gives the vector form for the pressure force in the equation above.

For studying the large-scale flow in planetary atmospheres, only the horizontal component of the Coriolis force matters. By large-scale we mean that the horizontal scale L_h of the motion is much larger than the vertical scale L_v, e.g., the scale height. For $L_h \gg L_v$, the vertical winds are small compared to the horizontal winds, and the vertical accelerations are small compared to the horizontal accelerations. Both accelerations are small compared to gravity, but gravity only acts in the vertical direction. So we can neglect the vertical wind in the horizontal component of the equation of motion, and we can neglect all accelerations in the vertical component, which

(*Box 8.2 continued*)

simply becomes the equation of hydrostatic equilibrium. The component equations are then

$$\frac{du}{dt} - fv = -\frac{1}{\rho}\frac{\partial P}{\partial x}, \quad \frac{dv}{dt} + fu = -\frac{1}{\rho}\frac{\partial P}{\partial y}, \quad 0 = -\frac{1}{\rho}\frac{\partial P}{\partial z} - g$$

Here u is the eastward component of velocity and v is the northward component; x and y are the corresponding horizontal coordinates, and z is the vertical coordinate. $f = 2\Omega \sin\phi$ is the Coriolis parameter, where ϕ is latitude. The terms involving f are the only significant terms in the Coriolis acceleration, because we are only interested in the horizontal force balance, the vertical force balance being dominated by pressure and gravity. If we put the Coriolis terms on the right side of the first equation, $+fv$ is a force per unit mass to the east, and it arises from a wind to the north. In the northern hemisphere where f is positive, this force is to the right of the wind. Similarly, $-fu$ in the second equation is a force per unit mass to the north, and it arises from a wind to the west (the minus x direction). Again the Coriolis force is to the right of the wind in the northern hemisphere. The Coriolis force (actually its horizontal component, which is what we care about) vanishes at the equator. In the southern hemisphere, where f is negative, the Coriolis force is to the left of the wind.

low pressure is on the left, so the net force—the pressure gradient force—is to the left. The balance is called geostrophic balance (box 8.4). Except at the equator, large-scale, slowly rotating jets and vortices are always in approximate geostrophic balance. Even if the current is

Box 8.3.
Hockey on the Merry-go-round

To visualize the inertial forces in a rotating system, imagine a hockey puck sliding without friction on a rink that was frozen to a merry-go-round. The water was allowed to freeze before the merry-go-round was turning, so the surface of the ice is flat. Then, with the merry-go-round turning counterclockwise, we go out some distance from the center and put the puck on the ice. Viewed from outside, the puck was moving with us in a circle, so when we released the puck, it continued on a line tangent to the circle, eventually sliding off the edge. Viewed from the ice, the puck was stationary, but we were pulling it toward the center to balance the centrifugal force. Facing outward, we saw the puck starting to move radially away from us, but then it lagged behind—it drifted to the right as we continued rotating with the merry-go-round to the left. The initial outward movement was due to the centrifugal force, but the deflection to the right was due to the Coriolis force.

Now let the rink freeze while the merry-go-round is rotating. The surface of the ice is curved upward into a parabolic bowl shape, and it is an equilibrium surface. We don't have to keep our hand on the puck to prevent it from sliding off. When we release it, the puck doesn't move. The resultant of the centrifugal force and gravity now acts perpendicular to the surface, so the puck is in equilibrium. If we disturb the equilibrium by releasing the puck with an initial sliding velocity, the Coriolis force will kick in. Viewed from outside, the puck will travel in a straight line. However, the merry-go-round is rotating counterclockwise, and viewed from the ice, the puck's velocity vector is rotating clockwise. We on the ice would say that there was a Coriolis force acting to the right of the velocity.

(*Box 8.3 continued*)

In the rotating reference frame the puck moves clockwise in a circle. If it is moving in a straight line in the rotating reference frame, there must be another force acting to the left of the velocity to balance the Coriolis force.

Box 8.4.
Geostrophic Balance

Two inequalities are important for the dynamics of large-scale flows in rotating systems. The inequality $L_h \gg L_v$ leads to the hydrostatic approximation (box 8.2), which is a balance between gravity and the vertical pressure gradient. The inequality $U/(fL_h) \ll 1$ leads to the geostrophic approximation, which is a balance between the Coriolis acceleration and the horizontal pressure gradient. Here U is the typical magnitude of the horizontal velocity, and U/L_h is the rate at which a flow feature moves its own diameter. U/L_h is a measure of the time derivative in the component equations of box 8.2. The time derivative of the velocity is of order U^2/L_h, and the Coriolis term is of order fU. The ratio of the two terms is $U/(fL_h)$. This ratio is called the Rossby number, after Carl-Gustaf Rossby, a meteorologist of the early twentieth century. If the Rossby number is small enough, say 0.1 or less, then the time derivative terms can be neglected and the component equations become

$$-fv = -\frac{1}{\rho}\frac{\partial P}{\partial x}, \quad fu = -\frac{1}{\rho}\frac{\partial P}{\partial y}, \quad 0 = -\frac{1}{\rho}\frac{\partial P}{\partial z} - g$$

These are the leading order terms. There are no time derivatives in these equations; they enter at the next level of the approximation. The leading order terms express a balance between Coriolis accelerations and pressure gradient forces. Instead of Coriolis forces making the flow move clockwise in a circle in the northern hemisphere (box 8.3), the flow can move in any direction as long as there is a pressure force to balance it. In the northern hemisphere ($f > 0$) the high pressure is on the right of the wind and the low pressure is on the left, as can be seen by examining either of the horizontal component equations.

At the equator, the Rossby number cannot be small and geostrophic balance cannot apply, because f vanishes there. Rotation dominates at high latitudes and becomes less important at low latitudes. This transition has important consequences: For instance, symmetric circulations like the Hadley cell dominate at low latitudes and eddy circulations dominate at high latitudes. The latitude of the transition is different for different planets and is an important parameter in Earth's climate.

---

not exactly in a straight line, the high pressure is on the right. This is why the flow is counterclockwise around a low-pressure center in the northern hemisphere and is clockwise around a high-pressure center. The directions are reversed in the southern hemisphere.

Why are Coriolis forces so important for the weather of rotating planets and so unimportant in daily life? The answer has to do with the time scale of events in our daily lives. A batter hitting a 95-mile-per-hour fastball has 0.43 seconds from the time the ball leaves the pitcher's hand to the time the ball crosses the plate. During that time, the Earth has rotated 0.0017 degrees. The line between the pitcher and the catcher has rotated by the same amount—the catcher moving to the left as seen from the pitcher's mound, assuming the game is being played at midlatitudes in the northern hemisphere. The ball travels in a straight line relative to a nonrotating reference frame, but it appears to curve to the right in the northern hemisphere. The result is that the ball will miss its target by 0.011 of an inch. Contrast this small effect with that on a storm system that rotates once in several days around its axis, or a current that flows for days on end in a straight line relative to the surface of a rotating planet. Without the pressure gradient force pushing to the left, the deflection of the wind by the Coriolis force would be huge.

The Rossby number $U/(fL_h)$ is an inverse measure of the importance of rotation. U/L_h is the rate at which a vortex moves its own diameter, or it is the velocity of the ball divided by the distance to the plate. If that rate is slow compared to the rate at which the planet turns, then the Rossby number is small and rotation is dominant. For the pitcher and catcher, the Rossby number is 10^4 and rotation is negligible. For a storm of diameter 2000 km and wind speeds of 10 m s^{-1}, the Rossby number is 0.05. Thus large-scale, slow-moving flows are dominated by rotation. Since U and L_h are vague terms ("typical" velocities and length scales), there is no unique definition of the Rossby number. Nevertheless, it provides order-of-magnitude estimates of terms in the equations of motion.

Geostrophic balance implies that the high pressure is to the right of the flow in the northern hemisphere and to the left of the flow in the southern hemisphere. Equivalently, one can say that the height of a constant pressure surface is greater to the right of the flow in the northern hemisphere and greater to the left of the flow in the southern hemisphere. This is because pressure increases with depth, so greater height means greater pressure underneath (box 8.5). Examples include the midlatitude jet streams on Earth, which blow generally from west to east. The height of the 500-mbar surface is 5800 m near the equator and 5100 m at the pole. The eastward winds in between are in approximate geostrophic balance.

Geostrophic balance also explains why the speed of the jet stream increases with altitude from Earth's surface up to 10 km altitude. It is because the equator is warmer than the poles, and the air columns at the equator are taller than those at the poles. As a result, the height difference between the equator and poles is more pronounced at high altitudes than at low altitudes. The eastward wind is proportional to the height difference, so the eastward wind increases with height. This increase of eastward zonal wind with altitude when temperature decreases toward the pole is called a thermal wind (box 8.6). A warm-core vortex like a hurricane is another example. Standing at the center of a northern-hemisphere hurricane looking outward is like standing at the equator looking toward the pole. In both cases the warm air is at your back. The warm-core vortex becomes more clockwise with altitude, which is like the eastward zonal wind increasing with altitude.

Box 8.5.

Pressure Coordinates

Meteorologists use the height of constant pressure surfaces in describing the horizontal pressure gradient force (boxes 8.2 and 8.4). Weather maps show the height in meters of the 500 mbar surface, and general circulation models (GCMs) use pressure as a vertical coordinate, with height $z(x, y, P)$ as a dependent variable. This reverses the role of z and P compared to what appears in boxes 8.2 and 8.4. The transformation uses a mathematical identity

$$dz(x, P) = \left(\frac{\partial z}{\partial x}\right)_P dx + \left(\frac{\partial z}{\partial P}\right)_x dP$$

So if $dz = 0$, we have an expression for $(\partial P/\partial x)_z$.

$$\left(\frac{\partial P}{\partial x}\right)_z = -\frac{(\partial z/\partial x)_P}{(\partial z/\partial P)_x}, \quad \left(\frac{\partial P}{\partial y}\right)_z = -\frac{(\partial z/\partial y)_P}{(\partial z/\partial P)_y}$$

In both of these equations, the denominator is the upside-down version of the hydrostatic equation, $\partial z/\partial P = -1/(\rho g)$. Putting this into the geostrophic equations and treating $z(x, y, P)$ as the dependent variable yields

$$-fv = -g\left(\frac{\partial z}{\partial x}\right)_P, \quad fu = -g\left(\frac{\partial z}{\partial y}\right)_P, \quad 0 = -g\left(\frac{\partial z}{\partial P}\right)_{x,y} - \frac{R_g T}{P}$$

We used the equation of state $1/\rho = R_g T/P$ to eliminate ρ in the last equation. The minus signs are still there, so high $z(x, y, P)$ is to the right of the wind in the northern hemisphere. The statement has to be interpreted as high height of a constant pressure surface. When a constant pressure surface is high, the pressure is high, and vice versa. These equations allow one to compute u, v, and all the thermodynamic variables from the derivatives of the function $z(x, y, P)$.

Box 8.6.
Thermal Wind Equation

Starting from the geostrophic equations with $z(x, y, P)$ as the dependent variable (box 8.5), one can take $\partial/\partial P$ of the first equation and $\partial/\partial x$ of the third equation and subtract the result. Do a similar thing for the second and third equations and one obtains the so-called thermal wind equations

$$-f\left(\frac{\partial v}{\partial P}\right)_{x,y} = \frac{R_g}{P}\left(\frac{\partial T}{\partial x}\right)_P, \qquad f\left(\frac{\partial u}{\partial P}\right)_{x,y} = \frac{R_g}{P}\left(\frac{\partial T}{\partial y}\right)_P$$

One can use the hydrostatic equation in the form $dP = -\rho g dz$ and the equation of state $P = \rho R_g T$ to change back into height coordinates:

$$f\left(\frac{\partial v}{\partial z}\right)_{x,y} = \frac{g}{T}\left(\frac{\partial T}{\partial x}\right)_P, \qquad -f\left(\frac{\partial u}{\partial z}\right)_{x,y} = \frac{g}{T}\left(\frac{\partial T}{\partial y}\right)_P,$$

These two equations say the same thing, but with the x-y coordinates rotated through $90°$. For example, take the second equation in the northern hemisphere where $f > 0$. The equation says that the eastward velocity u increases with height when the temperature decreases with latitude on a constant pressure surface. In the southern hemisphere f is negative, but in each hemisphere if the temperature decreases toward the pole, the zonal wind will increase with height. Thus the thermal wind equation accounts for the midlatitude jet streams, which blow to the east in each hemisphere. In the northern hemisphere, if one is looking toward lower temperatures, the wind component to the right will increase with altitude.

As another example, consider a warm-core vortex like a northern hemisphere hurricane. The circulation is counterclockwise at the ocean surface, so the flow is to the left if one is looking outward from the center. Since the center is warm,

---

(*Box 8.6 continued*)

the thermal wind equation says that the wind to the right increases with altitude. This is equivalent to the wind to the left decreasing with altitude. The circulation may reverse at some intermediate altitude and become clockwise at high altitude. This is normally what happens in a terrestrial hurricane.

The thermal wind equations also say that when f goes to zero, the horizontal temperature gradients go to zero. Places where f is small include slowly rotating planets and the equators of rotating planets. At these places, the horizontal temperature gradients are small. They are not necessarily zero because other terms, which were negligible before, come into play. Temperature gradients are small in the Earth's tropics, and the equator-to-pole temperature gradients are small on Venus. According to this reasoning, Venus is like the Earth's tropics. The Coriolis force is weak, so the pressure gradients have less to push against, and they are weak. And since pressure and temperature gradients are related through the equation of state and hydrostatic balance, the temperature gradients are weak as well.

In the southern hemisphere the eastward wind also increases with altitude and the temperature decreases toward the pole, but here the temperature is increasing with latitude and the Coriolis force is to the left of the wind. The net result is the same: There is an eastward jet stream in each hemisphere whose speed increases with altitude, and both are in approximate geostrophic balance (box 8.6) with temperature greater at the equator and less at the pole. Similarly, hurricanes in the southern hemisphere become more counterclockwise with altitude.

Vorticity is a measure of spin, which is ubiquitous in rotating fluids (box 8.7). The sign of vorticity follows a right-hand rule, in which counterclockwise rotation is defined as positive and clockwise rotation as negative. For a patch of fluid rotating counterclockwise as if it were a solid body, vorticity is twice the angular velocity. For an eastward zonal flow u that depends on the northward coordinate y, the vorticity is $-du/dy$. Cyclonic vorticity is defined to be in the same direction as the vertical component of the planet's rotation. On Earth a large-scale cyclone is counterclockwise in the northern hemisphere and clockwise in the southern hemisphere. An anticyclone is the opposite. For all planets, the north pole is defined by the orbital motion around the Sun and not by the spin. So we say that Venus spins backward. Nevertheless, on all planets, a cyclonic vortex is in the same direction as the vertical component of the planet's rotation and is always a low-pressure center. Cyclonic and anticyclonic shear zones are bands of low and high pressure, respectively, that wrap around the planet at constant latitude.

The vorticity of the solid planet is 2Ω, and the vertical component is $2\Omega \sin\phi$, where ϕ is latitude. The quantity $2\Omega \sin\phi$ is known either as the planetary vorticity (box 8.8) or as the Coriolis parameter (box 8.2), depending on the context, and is represented by the symbol f. Planetary vorticity is an important term in the expression for potential vorticity (box 8.8), which is conserved on fluid parcels as they move and evolve. The derivative of f with respect to the northward coordinate y is called β and is equal to $(2\Omega/R)\cos\phi$, where R is the planetary radius. β

Box 8.7.
Vorticity

Cyclonic vorticity is defined to be in the same direction as the vertical component of the planet's rotation—counterclockwise in the northern hemispheres of Earth, Mars, Jupiter, Saturn, and Neptune, but clockwise in the northern hemispheres of Venus and Uranus. Venus and Uranus are different because they spin backward. The International Astronomical Union (IAU) defines north as the spin pole that most nearly coincides with the orbit pole. Therefore north on Venus and Uranus is up, roughly parallel to north on Earth, but the spin is backward. The alternative was to say that Venus and Uranus are upside down, but the IAU chose not to go there.

Since the high pressure is on the right for geostrophic flow (box 8.4), a low-pressure center is cyclonic and a high-pressure center is anticyclonic. This holds true both in the northern hemisphere and in the southern hemisphere, although the direction of circulation—clockwise or counterclockwise is different. Constant-pressure surfaces are high when the pressure is high (box 8.5), so a depression in the height is cyclonic and a bulge is anticyclonic. Finally, the thermal wind equation says that a warm-core vortex becomes more anticyclonic with altitude, both in the northern and the southern hemispheres (box 8.6). A cold-core vortex does the opposite.

Box 8.8.
Potential Vorticity

Potenital vorticty is a vorticity-like quantity that is also a conserved tracer, meaning that a fluid parcel keeps its value of PV as it moves around. Its PV changes only if it mixes with other parcels or loses kinetic energy through friction. Since these processes happen slowly compared to dynamical processes like winds circulating around a vortex, PV is a good tracer of the motion. In this sense, PV is like paint—it drifts with the fluid and keeps its color until the parcel mixes. However, paint is a passive tracer—it is not related to any dynamic variables and it doesn't affect the flow. PV does both, so it is an active tracer. Conservation of PV is related to conservation of angular momentum, which is the product of moment of inertia and angular velocity. A simple expression for PV is

$$\text{Potential vorticity} \ = PV = \frac{\zeta + f}{h}$$

Think of a thin disk of thickness h draped like a pancake on the surface of a rotating sphere. To simulate a large-scale weather system, the pancake is 1000 km in diameter and 10 km thick. It can only spin about a vertical axis. It cannot leave the surface of the sphere because it is held there by gravity. The relative vorticity, which is twice the spin due to its rotation relative to the planet, is ζ. The pancake also has vorticity due to the spin of the planet. The component around a vertical axis is the Coriolis parameter $f = 2\Omega \sin\phi$ (box 8.1). This is twice the component of the planet's angular velocity about a vertical axis. In this context, f is called the planetary vorticity and $(\zeta + f)$ is called the absolute vorticity, because it is the vertical component of vorticity in an absolute reference frame. The moment of inertia of a disk is (1/2)

(*Box 8.8 continued*)

MR^2, where M is the mass and R is the radius of the disk. The volume is $\pi R^2 h$, so MR^2 is inversely proportional to the thickness h if mass and density are constants. So with moment of inertia proportional to $1/h$ and angular velocity proportional to $(\zeta + f)$, conservation of angular momentum implies $(\zeta + f)/h$ is constant. Other expressions apply when density is not constant, but the notion of PV is perfectly general.

---

is known as the planetary vorticity gradient, and it figures prominently in the theory of planetary waves and in the stability of zonal jets (box 8.9).

8.3 ORDER FROM CHAOS

Perhaps we shouldn't have been surprised when Voyager revealed the chaotic small-scale motions near the Red Spot and other long-lived ovals. Dynamical models of Jupiter's atmosphere, based on the same principles that were used to build weather-forecast models for Earth, had not yet produced Red Spots in 1979, but they had demonstrated that ordered flow could evolve from chaotic motions. The process involves an inverse cascade of energy from small scales to large scales. The models considered a thin layer of fluid supported on the surface of a rotating planet.[91,92] The Coriolis force and the pressure force defined the laws of motion for this simple system. The flow was forced by continuously adding small vortices—spinning parcels of fluid, which were chosen

Box 8.9.

Waves and Stability

Conservation of PV helps us understand the dynamics of fluids on rotating planets. We will consider two examples. Both hinge on the increase of f with latitude, since f is proportional to $\sin\phi$. First consider a parcel moving to the east as part of a zonal flow. Let h be a constant. If the parcel takes a small excursion to the north, its f will increase and its ζ will have to decrease in order to conserve $(\zeta + f)$. If it started with $\zeta = 0$, its relative vorticity becomes negative. The parcel's rotation then is clockwise, and if the parcel is generally going east, its clockwise rotation will send it back to its original latitude. Similarly, a southward excursion gives it counterclockwise rotation, which also sends it back to its original latitude. In other words, the variation of f with latitude provides a restoring force for waves on an eastward current. The waves are called Rossby waves, or planetary waves. The variation of f with the northward coordinate y is called β (beta), where

$$\beta = df/dy = (2\Omega/R)\cos\phi$$

Here ϕ is the latitude, Ω is the rotation rate, and R is the radius of the planet. Rossby waves are stationary in an eastward zonal flow because of the beta effect. Since f is the planetary vorticity, it follows that β is the planetary vorticity gradient.

The other example involves the stability of a zonal flow. Again let h be a constant. A theorem due to Lord Rayleigh and modified by H-L Kuo says that a zonal flow $u(y)$ is stable if PV increases monotonically with latitude. Since ζ is $-u_y$, where subscript y is the derivative, the derivative of ζ is $-u_{yy}$, which is minus the curvature of the zonal velocity profile. And since the derivative of f is β, the theorem says that the flow is stable if $\beta - u_{yy} \geq 0$. This is called the Rayleigh-Kuo stability theorem. Since β

at random to be clockwise or counterclockwise. The kinetic energy put in by the forcing was removed by friction, and the flow was allowed to evolve.

In these numerical experiments, one finds that the like-signed vortices merge with one another to make larger vortices. This merging continues until the vortices reach a characteristic size $(U/\beta)^{1/2}$. Here U is the average speed of the flow, and β is the planetary vorticity gradient (box 8.9). When the vortices get close to this size, they stop growing in the north-south direction. They continue to grow in the east-west direction, stretching out until they become bands that wrap around the planet on constant latitude circles. The remnants of the vortices are the shear zones pictured in figure 7.6. The currents around their peripheries have become the zonal jets. The shear zones with an eastward current on the south side and a westward current on the north side, like the band that holds the Great Red Spot, have counterclockwise shear. These alternate in latitude with the clockwise shear zones, and their widths are of order $(U/\beta)^{1/2}$. If one regards the random input of small vortices as chaos and the final pattern of wide bands as order, then order has evolved from chaos. Certainly the scale of the flow has increased, and the direction of the flow has acquired an east-west orientation.

The small-scale eddies constantly putting energy into the zonal jets is another example of a negative viscosity phenomenon, which we discussed in connection with the superrotation of the Venus atmosphere (box 3.14). It seems to occur on Jupiter as well. A fluid parcel moving from southwest to northeast (SW-NE) on a diagonal trajectory is carrying eastward momentum northward. If it returns on the same SW-NE diagonal trajectory, it is carrying westward momentum southward. To do so, it must have deposited eastward momentum during its northmost excursion and deposited westward momentum during its southmost excursion. In both cases, the product of the northward eddy wind v' and the eastward eddy wind u' is positive. The average with respect to longitude is $\overline{u'v'}$, and it is the northward transport of eastward momentum per unit mass (box 3.14). If $\overline{u'v'}$ is positive, and if there is an eastward jet to the north and a westward jet to the south, then the eddies are accelerating the jets. More generally, if $\overline{u'v'}$ and $d\bar{u}/dy$ have the same sign, then the eddies are putting energy into the jets. Here $\bar{u}$ is the average eastward wind speed and

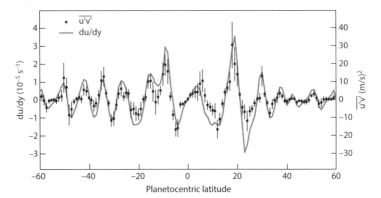

Figure 8.2. Eddy momentum transport vs. latitude on Jupiter. The continuous curve (*left scale*) shows the shear in the zonal wind. Where the shear is positive, the speed of the eastward wind is increasing with latitude. The points (*right scale*) show the northward transport of eastward momentum by the eddies. The positive correlation between the points and the curve means that the eddies are tending to increase the energy of the zonal wind.

(Adapted from fig. 5 of Salyk et al. [2005])[93]

y is the northward coordinate. If the signs are opposite, then the eddies are taking energy out of the jets. One can examine the eddies and jets on Jupiter to see which way they are arranged.[93] The result is shown in figure 8.2. The shear changes sign as one moves from one latitude to another, but when $\overline{u'v'}$ is positive, $d\bar{u}/dy$ is positive as well. This means that the eddies are putting energy into the jets and not the other way around.

The multiple jet streams demonstrate a striking difference between the weather patterns on Jupiter and those on the Earth. On Earth there is one eastward jet in each hemisphere. On Jupiter there are five or six (fig. 7.6). What's more, the jet streams are remarkably steady. They don't meander north and south, as jet streams on Earth do. The Jovian jets stay at their chosen latitudes unless they are interrupted by a large oval like the Great Red Spot. Even these interruptions are very stable and predictable, since both the jets and the ovals are constant in time. Changes in the speeds and latitudes from the Voyager encounter in 1979 and the Cassini encounter in 2000 are almost undetectable (fig. 8.1).

The numerical models of jets forced by eddies lead to jets whose characteristic width is of order $(U/\beta)^{1/2}$. Another way to look at this is through the stability of the jets—why don't they break up into waves and eddies (box 8.9)? A zonal flow may be steady, but it is not necessarily stable. Steady means that the flow doesn't vary with time. Stable means that the flow doesn't vary even if you perturb it slightly. An example is a zonal flow with a small-amplitude, wavelike perturbation. Waves come in a variety of shapes and sizes (wavelengths). If none of the physically possible waves can grow, then the flow is stable. If any of the waves can grow, then the flow is unstable. Usually stability depends on the gradients of the flow and not on the flow itself. For an atmosphere on a rotating planet, the simplest criterion is that the flow is stable if $\beta - \bar{u}_{yy} \geq 0$, where β is the planetary vorticity gradient (box 8.9) and $\bar{u}_{yy}$ is the curvature of the velocity profile—the second derivative with respect to the northward coordinate y. This is the so-called Rayleigh-Kuo stability criterion or sometimes the barotropic stability criterion (box 8.9). Since β is positive at all latitudes, the stability criterion is most likely to be violated at the centers of the westward jets, where $\bar{u}_{yy}$ is large and positive. Comparison with observation is difficult because taking the second derivative of noisy data is fraught with uncertainty. But one can draw parabolas of the form $u = u_0 + \frac{1}{2}\beta y^2$ and plot them alongside the measured velocity profile $u(y)$. The result is shown in figure 8.3.[94] Many of the westward jets have curvature equal to that of the parabola, which means that their width in latitude is of order $(U/\beta)^{1/2}$ This generally agrees with the numerical models, but some of the westward jets have more curvature than the parabola, indicating that the flow violates the barotropic stability criterion.

Violation of the stability criterion does not necessarily mean that the flow is unstable, but for many types of flow it does. The numerical models,[90,91,92] in which small-scale eddies are constantly putting energy into the zonal jets, never violate the stability criterion, although they come close. The fact that the width of the jets is of order $(U/\beta)^{1/2}$ means that β is controlling the curvature of the jets, and this is consistent with the numerical models. But the fact that the curvature of the westward jets sometimes exceeds β is inconsistent with the numerical models, and this is a mystery. This relation between the width of the jets and their speed generally agrees with observation, but it does not establish the value of either width or

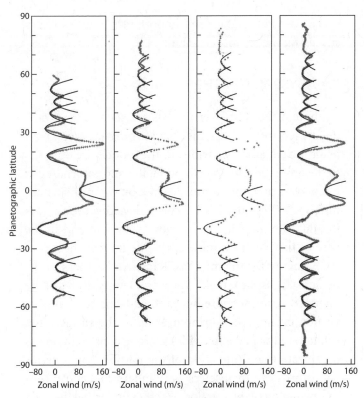

Figure 8.3. Application of the barotropic stability criterion to Jupiter's zonal jets. The theorem says that the flow is stable if the curvature of the zonal velocity profile (second derivative with respect to the northward coordinate) is less than the planetary constant β. The points are measurements of the zonal velocity profile, and the parabolas have curvature equal to β. The jets generally follow the parabolas, indicating that the beta effect is relevant. But the fact that some of the points fit within the parabolas means that those jets violate the criterion and could be unstable.

(Adapted from fig. 8 of Li et al., 2004)[94]

speed. Obviously, these are important quantities that a successful theory must explain.

8.4 ABYSSAL WEATHER

Although the thin-layer models are successful in producing jets from eddies, they tend to produce a westward jet at the equator. The winds in the models are opposite to the observed zonal winds on Jupiter and Saturn. A different class of models simulates motions in the planetary interior.[95] These models had been developed for studying fluid motions in the interior of the Sun and in the core of the Earth. In these cases, the internal heat provides energy for the motions. Radiative heat loss in the planet's atmosphere makes the warm temperatures of the planetary interior more buoyant, and this sets up convection currents that bring heat up to the surface. The rotation of the planet provides a strong constraint on the organization and direction of the motion. The interior winds organize themselves as cylinders concentric with the planet's axis of rotation. Each cylinder has its own angular velocity. These models suggest that the zonal velocities we see in the atmosphere (fig. 7.6) are the surface manifestation of the rotation on cylinders in the interior.

Interestingly, in these models the outermost cylinders, which intersect the surface at low latitudes, rotate faster than the average. In other words, the deep models give eastward jets at the equator, as observed. The

Galileo probe entered the atmosphere at 7° north latitude. It was tracked down to the 20 bar level, which is
only 0.2% down to the center of the planet starting at the
tops of the clouds. But in that layer, about 150 km thick,
the winds increased with depth from 100 m s⁻¹ at cloud
top to 180 m s⁻¹ at 20 bars.[96] This does not prove that the
winds extend through the planet on cylinders, but it provides support for the theory that the winds are deep. At
that one place on the planet, the winds below the clouds
were greater than the winds within the clouds.

These successes of the interior models do not mean
the issue is settled. First, on Uranus and Neptune, cloud
features at the equator move in the retrograde direction,
or slower than the planet as a whole. Any valid theory
has to explain the winds on all four giant planets. Second,
the parameters of the models affect the winds. Several
modeling groups are trying to treat the eddies realisti-
cally, letting them draw energy from sunlight heating the
equator and from internal heat rising from the interior.
Other groups have been experimenting with the viscosity, letting it be a function of latitude to reflect the greater
penetration depth of the cylinders at high latitude into
the planetary interior.[97] Changing these parameters affects the direction of the jets, and some thin-layer models
give eastward jets at the equator.

The variation of the flow with depth is a major unknown, and it adds uncertainty to the models. A thin-
layer model is a good approximation in a region of
strong, stable stratification, where the lapse rate is much

less than the adiabatic lapse rate (box 3.2). The stable stratification inhibits vertical motion, and the flow is approximately horizontal. Stable stratification is likely within and above the clouds, but it becomes less likely below cloud base. With weak stratification, the vertical component of the flow can be as large as the horizontal component, and that invalidates the thin-layer models.

Currently there is a disconnect between the two kinds of models. One problem is lack of information, and the other is lack of resources. Although we have exquisite information about the flow at the tops of the clouds, we have very little information about the flow structure below. One can build models of the deep flow, but it is hard to test them. The resource problem is that no computer can handle thin-layer models and 3-D models simultaneously. The thin-layer models require high spatial resolution and small time steps, and the 3-D models require long numerical integrations. The compromise is that thin-layer models are run with simplified assumptions about the flow underneath, and 3-D models are run with unrealistically coarse resolution.

One hope is that spacecraft will be able to measure the signature of winds in the interiors of Jupiter and Saturn. The Coriolis forces associated with the winds are balanced by pressure forces, which require rearrangements of mass in the planetary interior. If the winds are deep enough and the mass involved is large enough, the effects will show up as slight irregularities in the gravity field. The Juno spacecraft will measure the gravity field of Jupiter, and the Cassini spacecraft will measure the gravity field of Saturn, both in 2017. The spacecraft will be in polar orbits that bring the spacecraft close to the planet. The perturbations on the orbits of the spacecraft will reveal the deep winds if they are present. Also, Juno with its radio eyes can peer through the clouds down to pressures of 50 bars or more, and can see whether there are variations of water and ammonia at those levels. This will tell us if there is weather down there, or whether it is a chemically uniform, hot, dark, infinite abyss with no weather whatsoever.

8.5 ZONES AND BELTS

Until the mid-twentieth century, amateur astronomers were the authorities on the spots and bands in Jupiter's atmosphere. Most of the professional astronomers were busy studying stars and galaxies, and most of the professional meteorologists were studying Earth. The amateurs called the light and dark bands zones and belts, respectively. Although the zones and belts changed contrast and sometimes appeared to merge with their neighbors, they always reappeared at the same latitudes (fig. 8.1). They never disappeared forever. Consequently, they acquired names. Here is the sequence in the northern hemisphere: Equatorial Zone, North Equatorial Belt, North Tropical Zone, North Temperate Belt, North Temperate Zone, North North (*sic*) Temperate Belt, and so on. Spots circle the planet on lines of constant latitude, and the amateurs timed the spots' recurrence periods at each latitude. Since the internal rotation period was unknown, the amateurs defined a uniformly rotating reference frame for measuring spot positions. The observations revealed that the jets are on the boundaries between the belts and the zones, with the westward jets on the poleward sides of the belts and the eastward jets on the poleward sides of the zones. Thus the belts are cyclonic shear zones and the zones are anticyclonic shear zones. The jets are the most permanent feature of the belts and zones. The colors and contrast change more than the jets. The Great Red Spot sits in the South Tropical Zone, although it protrudes into the South Equatorial Belt to the north. In general, the long-lived oval spots are anticyclones and they sit in the zones, which are also anticyclonic.

We do not fully understand the connection between the colors of the belts and zones and their dynamics. We don't know why the Red Spot is red, for instance. Color is like paint, and it depends on chemistry. If it were a strictly passive tracer, then it would mix to a uniformly bland color. Obviously color is not just a passive tracer—the chemistry must be responding to the dynamics because the dynamical structures like the belts and zones and long-lived ovals each has its own character-istic color. Understanding the chemistry would help us understand the dynamics and vice versa, but we are not at that point yet. The atmospheric gases are transparent, so the colors must originate on the cloud particles. The coloring agents could be sulfur, phosphorous, or organ-ics (carbon compounds). Identifying the coloring agents is difficult because solid materials have broad spectral features that cannot be uniquely identified. Gases are the opposite. Their spectra have narrow lines that give each gas a unique fingerprint (box 3.11).

The amateur astronomers, with their moderate-sized telescopes, have an important role to play. With high-speed digital cameras and the ability to rapidly process thousands of images, the amateurs now can take images of exceptional clarity. Moreover, they can respond to interesting weather phenomena on the giant planets in less than a day. In contrast, the large telescopes on Earth and the instruments on planetary spacecraft are scheduled weeks or months in advance and are difficult to reschedule. Thus the professional astronomers and the amateur astronomers complement each other.

The professional astronomers have access to high-resolution spectrometers and sensitive infrared detectors. Before the first spacecraft, they had measured the abundances of various gases—principally methane, ammonia, and hydrogen (box 3.11). They discovered that Jupiter has an internal energy source. And they determined the heights of the clouds in relation to the belts and zones. Figure 7.1 shows that 5-micron radiation is greatest in the belts and least in the zones. Recall that the gases in the atmosphere are transparent to 5-micron radiation but the clouds are not. The 5-micron hot spots are holes in the clouds where thermal emission from deeper, warmer levels can escape. Thus the zones are regions of uniform high clouds, and the belts are regions of patchy clouds with views to deeper levels. This led to the notion that the zones are sites of upwelling and the belts are sites of downwelling. The zones were supposed to be like the

Earth's intertropical convergence zone (ITCZ), a region of frequent thunderstorms where precipitation exceeds evaporation $(P-E > 0)$. The belts were supposed to be like the subtropical zones on either side of the equator, regions of fewer clouds and low relative humidity where $P-E < 0$. The circulation in the meridional plane (a vertical plane aligned north-south) was supposed to resemble the Hadley circulation on Earth (box 3.12), with low-level convergence in the zones and upper-level convergence in the belts. Whereas Earth had one ITCZ, Jupiter had several in each hemisphere, according to this view.

The idea of upwelling in the zones and downwelling in the belts got support from the ammonia and vorticity distributions.[98] The ammonia abundance is greatest in the zones and least in the belts, suggesting that the zones are moist updrafts and the belts are dry downdrafts. According to this hypothesis, the air loses its ammonia in the updrafts through precipitation and descends in the belts to pick up more ammonia deeper in the atmosphere. Further, the vorticity is anticyclonic in the zones and cyclonic in the belts. If one assumes that the vorticity is zero at cloud base—that the deep atmosphere is not moving except for small-scale turbulence—then the anticyclonic circulation of the zones would be like a warm-core vortex on Earth, where the winds become anticyclonic at high altitude. The cyclonic circulation of the belts would be like a cold-core vortex on Earth. So one would have hot air rising in the zones and cold air sinking in the belts, which is consistent with a thermally direct circulation like the Hadley circulation on Earth (box 3.12).

The Hadley cell has heating by latent heat release in the rising air and cooling by outgoing infrared radiation in the sinking air. That was the assumption for the belt-zone circulation on Jupiter. However, observations of lightning by the Galileo orbiter upset this assumption.[83] The orbiter observed lightning flashes on the night side and correlated their locations with clouds on the day side (fig. 7.4). Lightning is an indicator of moist convection, in which the air is heated by condensation of water vapor. The lightning clouds look like thunderstorm clusters on Earth with cloudless areas in between. The problem is that the lightning occurs in the belts,[99] which were supposed to be the sites of downwelling. The Hadley cell interpretation requires lightning and moist convection at the sites of upwelling, and those were supposed to be the zones. Also, the thunderstorm clusters probably require updrafts below the clouds to maintain the moisture supply. This led to speculations about two Hadley cells, one on top of the other, going in opposite directions.[100] In this view, the belts would have upwelling at low altitudes and downwelling at high altitudes. The zones would be the opposite. This hypothesis has not been verified in the numerical models. Part of the problem is that we do not have a good understanding of moist convection in giant planet atmospheres.

The vorticity of the belts is cyclonic, meaning that they are bands of low pressure. On Earth, areas of low pressure are areas of convergence, since friction with the surface causes winds in the atmospheric boundary layer to converge there. Over the ocean this low-level convergence brings in moisture, which leads to moist convection. A hurricane is the best example. Moist convection in the belts is consistent with lightning in the belts, but there is no atmospheric boundary layer to bring in the moisture. Perhaps convergence occurs below the base of the clouds anyway. The giant planets force us to think about what role land and ocean surfaces play in the dynamics of the atmosphere. Without such surfaces, the dynamics could be different from what they are on Earth.

Perhaps the whole picture of rising in the zones and sinking in the belts is wrong. The picture uses the thermally forced Hadley cell analogy, but on Earth the Hadley cell is confined to the tropics. The tropics are special because the Coriolis force is small there. On Venus and Titan the Hadley cell extends to higher latitudes because those planets are slow rotators and the Coriolis force is small everywhere. In contrast, Jupiter is a rapid rotator, and we should probably expect the Hadley cell to be confined closer to the equator than on Earth. In that case, the Hadley cell analogy should not apply to the multiple belts and zones at higher latitudes.

A better picture might be zonal jets forced by eddies (boxes 3.13 and 3.15), as discussed in section 8.3, "order from chaos." An eddy-driven circulation[12] might resemble the Earth's Ferrel cell. It is a thermally indirect circulation, in which hot air is forced down and cold air is forced up. It exists at midlatitudes as a residual; it is not the dominant weather feature at those latitudes. Instead, the midlatitude weather is dominated by the eddy heat flux (box 3.13), which overshoots and drives the Ferrel cell. Jupiter is different in this respect. On Jupiter the zonal winds are much larger than the eddy winds, and the belt-zone structure is the defining feature of the weather. What forces the Jovian eddies is still an open question. They could be like thunderstorms and hurricanes—forced by an energy source below the clouds, or they could be like midlatitude storms—forced by horizontal temperature gradients that arise from the uneven distribution of sunlight. If the zones and belts are like the Ferrel cell on Earth, one would have to explain all the other features—the uniform high clouds in the zones, the 5-micron hot spots in the belts, the lightning in the belts, and the abundant ammonia in the zones, which would have to follow from the processes that maintain the jets. These are some of the constraints that the correct theory must satisfy, and we are not there yet.

9.1 FATHER PLANET

IF VENUS AND EARTH ARE SISTER PLANETS, THEN Saturn[8] and Jupiter[7] should be brothers. The ancient Romans said they were father and son, and that is equally appropriate. They share many basic characteristics, but their climates are distinctly different. They both have multiple jet streams and banded cloud patterns. Saturn's weather is normally very calm, but every 20–30 years a giant storm erupts.[101] These storms last for a few months and then disappear. In contrast, Jupiter's giant storms endure without change for decades or centuries. Saturn's winds are stronger than Jupiter's although Saturn is twice as far from the Sun. In this chapter we first review the variables that might control the climate, and then we discuss how the climates actually differ.

Saturn and Jupiter are the largest and most massive planets in our solar system. Both planets are ~10 times larger than Earth: Saturn's equatorial radius at the level where P = 1 bar is 60,268 km, and Jupiter's is 71,492 km. Saturn's mass is 95.2 times Earth's mass, and Jupiter's mass is 317.8 times Earth's mass. Saturn is 9.5 AU from the Sun, and Jupiter is 5.2 AU from the Sun, where the AU is one-half the diameter of Earth's orbit around the

Sun. Saturn and Jupiter are basically big balls of hydrogen and helium gas. Because Saturn is less massive, its gravity is less and the gas is less compressed. At the 1-bar pressure level, which is just below the cloud tops, the gravitational acceleration g is about 10 m s^{-2}. That's the same as on Earth, whereas g on Jupiter is ~2.5 times Earth's value.

Saturn's internal heat is somewhat greater than what one would expect from an object with Saturn's mass and solar composition that has been cooling for 4.5 billion years. Since Saturn has 70% of the surface area and only 33% of the mass of Jupiter, it should cool faster. Its internal heat should now be less than what we observe. Apparently there is another source of internal energy besides simple cooling and the contraction that goes with it. Precipitation of helium raindrops[77] from liquid metallic hydrogen is the most likely source. We talked about this mechanism as it applies to Jupiter, but the effect is more pronounced with Saturn. Since Saturn cools faster, its interior is now cooler than Jupiter's, and more of the helium raindrops would have condensed out of the metallic hydrogen envelope. Condensation releases latent heat. In addition, the raindrops would have fallen down to the top of the rocky metallic core, which is estimated to occupy the inner 15–20% of the radial distance from the center of Saturn. The raindrops are heavier than the hydrogen they displace, so there is a net release of potential energy as they fall. The energy goes into heat when the raindrops slow down and stop, and that heat contributes to the internal energy still coming out of the planet today. Support for this theory comes from the He/H ratio in Saturn's atmosphere, which is only ~60% of the solar value. Jupiter's ratio is ~81%, as shown in table 7.1. For Saturn, the He/H measurement depends on estimating density, pressure, and temperature at the same level, from which the molecular mass is derived (box 3.1). It is an indirect measurement using remote sensing techniques, but the result has important implications for planetary evolution.

For climate, the important parameters are the energy supply, the temperatures, and the composition of the atmosphere. The ratio of internal heat to absorbed sunlight is 0.8 for Saturn and 0.7 for Jupiter, which means that the total radiated power relative to the solar power is 1.8 for Saturn and 1.7 for Jupiter. But since Saturn is farther from the Sun than Jupiter, the effective radiating temperature is 94 K for Saturn and 124 K for Jupiter. The power emitted by Saturn per unit surface area is 31% that of Jupiter and only 2% that of Earth. One might expect that the wind speeds would be proportionately lower and the storms would be proportionately weaker, but apparently that's not how atmospheres work.

There has never been an instrumented probe into Saturn's atmosphere, so we know much less about its composition than we do about Jupiter's. With one exception, we cannot confidently compare the enrichment factors for Saturn with those for Jupiter. The exception is carbon, which appears on both planets as methane. Methane is so volatile that it does not condense at any level in either planet's atmosphere. Therefore its abundance relative to hydrogen is likely to be constant throughout the atmosphere. If one can measure the CH_4/H_2 ratio at any altitude, one can make a fairly good estimate of the C/H ratio for the planet as a whole. Both CH_4 and H_2 leave their signatures in the spectrum of sunlight reflected by the planet and in the spectrum of infrared radiation emitted by the planet (box 3.11). The depth and breadth of the spectral lines are a measure of the amount of gas present. Uncertainties arise because methane and hydrogen absorb in different parts of the spectrum, and the photons take different paths through the atmosphere. Fortunately, the uncertainties are small enough to allow us to make some useful comparisons between Saturn and Jupiter.

Saturn seems to have a higher C/H ratio than Jupiter.[79] The corresponding enrichments relative to solar composition are ~9 and ~4.5, respectively. If these enrichments apply to other elements like O, N, and S, then Saturn has twice as many condensable gases and twice as much latent heat in its atmosphere as Jupiter. These elements would exist as vapors in the deep atmospheres below the base of the clouds as H_2O, NH_3, and H_2S. The assumption that the enrichment factor for C applies to the other elements gets some support from the Galileo probe results (table 7.1). One has to ignore the small variations between the enrichments of C, N, and S, and one has to treat the low O/H abundance as an accident of where the probe went in—an unusually dry spot in Jupiter's atmosphere.

There is much uncertainty about the water abundance on Jupiter, but other data also support the higher enrichment factors on Saturn. When one does a careful calculation of Saturn's interior structure assuming solar composition, taking into account the high-pressure equation of state for hydrogen-helium mixtures and the degree of self-compression due to Saturn's gravity, one gets a model planet whose density is too low. In other words, one needs a significant enrichment of heavy elements to match the observed density.[71] One cannot put the additional heavy elements into the core because then the model planet would be not oblate enough. The observed oblateness requires significant enrichment at all levels, farther from the axis of rotation, where the extra mass can be acted upon by the centrifugal force and contribute to the oblateness. This points to O, C, N, and S, which form gases that mix well with hydrogen and helium, rather than the abundant metals like Si, Mg, and Fe, which form metals and rocks that sink to the core. On the Sun, O, C, and N are more abundant than Si, Mg, and Fe, so it is unlikely that Saturn would be enriched only in the metals and not in the more volatile gases. The enrichment derived from density and oblateness is consistent with the enrichment derived from the C/H ratio in the atmosphere.

The latent heat is tied up in the condensable gases H_2O, NH_3, and H_2S. Latent heat is measured in units of energy per mole of condensing vapor (one mole = 6.02×10^{23} molecules). Multiplying by the fractional number of molecules, one gets the energy that would be released in a mole of atmosphere if all the vapors were to condense. Dividing by the heat capacity (energy per mole per degree Kelvin), one gets the corresponding temperature rise. Water is ~7 times more abundant than ammonia and hydrogen sulfide, and its latent heat per mole is almost twice as large. Therefore most of the contribution to the temperature rise is from water. For 9 times solar enrichment factors in a hydrogen-helium atmosphere, the temperature rise is 14 K, which is substantial for an atmosphere at 100 K. For Jupiter with a 4 times solar enrichment factor, the temperature rise is 6.2 K. In comparison, for saturated tropical air at 30 °C, the temperature rise due to complete condensation of water is 65 K. These are extreme conditions for Earth. The fractional temperature increase due to condensation, 14/100 for Saturn, is about two-thirds the 65/303 fractional temperature increase for Earth.

9.2 THE GREATEST STORM

Thus it appears that Saturn has two or three times as much latent heat, Joules per mole, as Jupiter. On the other hand, Saturn's atmosphere has only 31% of the power per unit surface area as Jupiter. This leads to a question: Do we expect Saturn's storms to be weaker than those on Jupiter because the power is less, or do we expect them to be more energetic because the stored latent heat is greater? In the latter case, the storms would have to be less frequent in time and more spread out in area, since the power per unit area is a fundamental limitation.

Apparently Saturn has chosen the second option. In December 2010, a giant lightning storm erupted on

Figure 9.1. Planet-encircling storm on Saturn. This picture was taken by the Cassini camera on February 25, 2011, about 12 weeks after the storm began. During that time the tail of the storm had wrapped around the planet and was encountering the head on the south side from the west.

(Source: http://photojournal.jpl.nasa.gov/, PIA12826)

Saturn,[102,103,104] It lasted for half a year and covered a band of latitude 10,000 km wide and 300,000 km long (fig. 9.1). Clouds from the storm had wrapped around the planet by early February 2011. The 300,000 km length is simply the circumference of the planet at the storm's 35° latitude. Planet-encircling storms happen every thirty years, on average, on Saturn.[101] The first recorded one was in 1876, and the one before the 2010–2011 storm was in 1990. The 2010–2011 storm was the sixth in recorded history and the first to occur with a spacecraft in orbit around the planet. Cassini's radio and plasma wave science (RPWS) instrument was the first to detect the storm, which signaled its existence on December 5, 2010, with intense radio bursts characteristic of electrostatic discharges, that is, of lightning. A few hours after the radio detection, the Cassini camera saw a cloud 1000 km in diameter that stood out from the usual bland atmosphere at visible wavelengths. It was a lucky observation. The preprogrammed observation sequence included one to two minutes of idle time every day to allow the spacecraft to stick to its schedule. The camera was instructed to take a picture whenever the idle time was not needed, and one of those opportunities occurred during the first few hours of the storm.

The camera's first opportunity to observe the storm in detail was a daylong observation on December 23–24. However, the RPWS announcement was accompanied by an appeal to the amateur astronomers[103] who followed the storm in December as it continued to erupt. The storm was located in the middle of a westward jet, and it moved to the west a little faster than the undisturbed atmosphere. The leading edge of the storm—the head—maintained a blunt aerodynamic shape pointing westward during the storm's six-month lifetime (fig. 9.1). The head remained the center of convective activity, as defined by high thick clouds, strong anticyclonic vorticity (box 8.7), and intense lightning activity. The RPWS detected lightning activity whenever the storm's head was above the radio horizon as seen from the spacecraft. This meant that lightning activity was grouped into daily episodes, each lasting for half the Saturnian day, which at the latitude of the storm was about ten hours and forty minutes long. Storm clouds that pushed into the neighboring currents to the north and south of the storm moved off to the east and eventually encountered the head in early February 2011. The body of the storm, at the same latitude as the head, slowly expanded to the east. A large anticyclonic vortex 10,000 km in diameter marked the eastern extent of the storm, not counting the currents to the north and south. When the large anticyclonic vortex had circled the planet and collided with the storm in mid-June, the head disappeared and the lightning activity slowed.[105] At this point the body of the storm stretched around the planet covering a latitude band ~10,000 km wide. During this final stage the disturbed clouds resembled one of the dark belts in Jupiter's atmosphere.

In the middle of each daily episode, the lightning flashes exceeded the ability of the RPWS to count them. The flashes overlap at rates greater than ten flashes per second, which is therefore a lower bound on the flash rate for the Saturn storm. The intensities of the radio waves are ~10,000 stronger than those of their terrestrial counterparts. The Cassini camera eventually saw the lightning flashes in day side images, but it was difficult. One problem is that the flashes do not look like the jagged bolts of lightning on Earth. Instead they are diffuse circular features ~200 km in diameter without a distinct edge. The inference is that the lightning is occurring below the cloud tops, and the photons are spreading out as they diffuse up from the lightning source. Puffy clouds illuminated from below by lightning are hard to distinguish from puffy clouds illuminated from above by sunlight. Night side images would have been helpful but none were in the sequence, which had been planned months before the storm began and was difficult to change. Nevertheless, the Cassini camera has detected visible lightning in the storm and has estimated its intensity.[106] The energy in visible light from single flashes on Saturn can be compared with those on Earth and Jupiter. The Galileo spacecraft mapped out the Jupiter flashes, but Voyager and Cassini detected them as well. The result is that the largest flashes are comparable on all three planets. This is a remarkable result, since the electrical properties of the atmospheres are so different and the storms that produce the lightning are different as well.

9.3 LIGHTNING AND MOIST CONVECTION

Lightning is a mysterious process even on Earth. How falling ice and liquid droplets cause the positive and negative charges to separate is not well understood. The mechanism probably has to do with the structure of water molecules—the positive charge resides on the two hydrogen atoms, which are crowded together on one side on the negatively charged oxygen atom. If such an asymmetry didn't exist, there might not be a systematic separation of positive and negative charge. On Earth, maximum charge separation seems to occur near the freezing level, where the larger ice particles falling through the cloud collide with smaller liquid droplets lofted on the updrafts. The net result is that positive charge is carried up and negative charge is carried down. As the charges separate, the electric field builds up until it reaches the breakdown value where the air ionizes. A huge electrical current flows through the ionized channel, down the path of least electrical resistance. This is the lightning, and it may jump from cloud to cloud or from cloud to ground. On the giant planets, the lightning must be from cloud to cloud.

Since the giant planets have three condensable gases in their atmospheres, it is possible that charge separation could occur at three different levels (fig. 7.2). If the enrichment factors for O, C, N, and S are the same, then water is the most abundant condensable gas (table 7.1). Unfortunately water would be undetectable at the tops of the clouds because the temperatures there are so low. If water were the most abundant condensable gas deeper down, then it would probably account for the most violent convective storms, the most rapid updrafts and downdrafts, and the greatest electrical charge separation. Unless the water abundance is extremely low, water has a freezing level where the ice and raindrops can interact. The same is not true of ammonia and hydrogen sulfide, which exist only in solid and vapor form on Jupiter and Saturn. Therefore, one expects the electrostatic

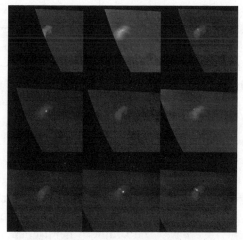

Figure 9.2. Lightning flashes on the night side of Saturn. The white cloud is ~3000 km long and is illuminated by light from the rings. Around each flash, the contour where the brightness is one-half the maximum value has a diameter of ~200 km. The duration of the sequence is 16 minutes. Exposure times were either 12 s or 120 s. This is the cover illustration for Dyudina et al. (2010).[106]

discharges to occur in the water cloud. There is observational evidence to support this statement (fig. 9.2).

The diameter of the visible flashes at the cloud top level is a rough indicator of the depth of the lightning. The farther the photons spread horizontally, the deeper is the inferred lightning flash. The ratio of the lightning depth to the flash diameter at cloud top depends on how the intervening clouds scatter the light, but it is roughly of order unity. This puts the lightning both for Jupiter and Saturn 125–250 km below the cloud tops, somewhere near the expected freezing level of water, although there is considerable uncertainty. The implication is that there is a level in these planets' atmospheres where one could find Earth-like conditions—warm temperatures, falling rain, and lightning.

A big difference between Jupiter and Saturn is the intermittency of lightning storms on Saturn. Saturn often has no lightning storms at all, judging by the absence of electrostatic discharges that the RPWS instrument would otherwise detect. Besides the great northern hemisphere storm of 2010–2011, there have been several lightning storms in the southern hemisphere.[106] They happen at intervals of one or two years, and they last for twenty-five to seventy-five days. The southern hemisphere storms never reached sizes greater than 2000 km in diameter, and they did not shed long tails. Instead, they produced trains of smaller spots that marched off to the west from the storm center. The storms were accompanied by lightning, which was observed by RPWS at radio wavelengths and by the Cassini camera at visible wavelengths. The southern hemisphere storms were an order of magnitude smaller than the northern storm in horizontal dimension and in time (duration of the storm). Saturn thus has less than one lightning storm, on average, at any one time. In contrast, Jupiter has about fifty lightning storms comparable in size to Saturn's southern storms on any given day, according to the Galileo orbiter results. Earth typically has two thousand thunderstorms spread around the planet on any given day.

Lightning arises from moist convection, which has an intrinsic property that leads to intermittency. In a dry atmosphere, convection occurs immediately whenever the lapse rate $-dT/dz$ exceeds the dry adiabatic lapse rate g/C_p. The upward heat transport makes dT/dz more positive (smaller lapse rate), which shuts off the convection (box 3.2). Potential energy never builds up, and the convection is a relatively mild process. Dust devils are the most vigorous form of dry convection. They acquire moderate amounts of potential energy because convection is inhibited close to the ground. In contrast, moist convection requires that the air be saturated before the convection begins. The moist adiabatic lapse rate is less than the dry adiabatic lapse rate, because condensation heats the parcel as it rises. An air column can have a lapse rate that is between these two adiabats. It can be stable to dry convection and unstable to moist convection. Still, nothing will happen until the air becomes saturated. This is called conditional instability.[11] It means that the potential energy can build up to a large value, but convection will not begin until the saturation condition is satisfied, at which point, one gets a large release of potential energy, as in a violent thunderstorm or tornado. If the air is nearly saturated, small updrafts can cause it to become saturated and can trigger a major moist convection event.

However, conditional instability might not apply to large-scale events like Saturn's giant storms. Certainly the storms are intermittent, which implies a slow buildup of potential energy that is suddenly released. This would imply a trigger, like a small updraft that causes the air to become saturated. However on Earth, the convective potential energy doesn't last for a long time. Threatening weather leads to a thunderstorm, usually in a few hours. Moreover, thunderstorms are basically local phenomena, each one covering an area ~10 km in diameter. On a larger scale, averaged over a day or more, moist convection is not that different from dry convection. It regulates itself by reducing the potential energy as fast as it is created. The Hadley cell is a good example: The individual thunderstorms come and go, but the large-scale pattern of air rising in the tropics and sinking in the subtropics is a steady, quasi-equilibrium feature of the general circulation.

Of course, moist convection on Saturn may be different from moist convection on Earth. On the giant planets, the rain has to evaporate beneath the clouds because there is no ocean or land surface to collect it. Also, with the flux of sunlight only 1% of the terrestrial value, it takes a hundred times longer to recharge the atmosphere after a big precipitation event. A third difference is that the mass of a water molecule (18 g per mole) is much greater than that of hydrogen (2 g per mole). This affects the buoyancy of air parcels. At a given pressure, the density of a saturated air parcel is proportional to m_{av}/T, where m_{av} is the average molecular mass of the mixture and T is the absolute temperature (box 3.1). The amount of water vapor is controlled by the saturation vapor pressure $e(T)$, which increases exponentially with T (box 5.1). At low temperatures, the parcel is mostly hydrogen, and density varies as $1/T$. At higher temperatures, the exponential in the numerator wins, and density varies roughly as $e(T)$. In other words, as temperature increases, the density of a saturated parcel first decreases and then increases. There is a density minimum in between.

This reversal in the density is similar, but in the opposite sense, to that of liquid water, which has a density maximum at T = 4 °C. Freshwater lakes that have been frozen all winter turn over in the spring as their surfaces warm up. In other words, convection occurs when heat is added at the surface instead of at the bottom, as is usual for convecting fluids. For Saturn, the analogous situation is when the cloud layer starts cooling after a large convective event. At first the density decreases while the clouds are dropping their load of water vapor. Later, when the air has dried out, further cooling leads to density increasing. While the density is low, it will be stable relative to the air below the clouds. Further cooling leads to instability and another large convective event.

The existence of a density minimum could lead to cyclic behavior somewhat like the turnover of freshwater lakes. Because the temperature at the tops of the clouds is so low, Saturn's cloud layer loses its heat slowly. It could take decades for the cloud layer to lose enough heat to start convecting again. To get this behavior, the mixing ratio H_2O/H_2 in the subcloud region has to be greater than 9×10^{-3}. At values below 9×10^{-3}, the exponential never wins; the cooling of the cloud layer would cause its density to increase, and convection with the subcloud layer would start immediately. There would be no decades-long wait for convection to begin. At values above 9×10^{-3}, cooling of the cloud layer initially shuts off the convection, and there is delay before the next convective event. A mixing ratio 9×10^{-3} corresponds to ~10 times solar abundance of water (table 7.1). Perhaps Jupiter has less enrichment than this critical value and Saturn has more. That difference could explain why Saturn has these giant storms and Jupiter does not.

Hydrogen atmospheres have other peculiarities. The spins of the two nuclei of the H_2 molecule can be parallel or antiparallel. One state of the molecule is called ortho and the other is called para. The difference in internal energies between the two states is substantial, especially at low temperatures. At temperatures below 100 K, conversion from one state to the other can lead to temperature changes of 40 K. The difference in internal energies means there is a difference in the specific heat, C_p, between the two states, which leads to a difference in the adiabatic lapse rates g/C_p. Since the conversion times between the two states are ten years or longer, potential energy can build up to substantial values before it is released in a convective event.[107] Simulations for Neptune yield multidecadal fluctuations in temperature and ortho/para fraction at the 1 bar level. However, simulations for Jupiter and Saturn do not exhibit such fluctuations, probably because their atmospheres are too warm.

Further modeling is needed to see if any of these mechanisms explain the giant Saturn storms. Some of the observables are: the 20–30 year time scale, the association with lightning and moist convection, the confinement in latitude, the propagation speed, and perturbations of the stratosphere above the storm.

9.4 SATURN'S ROTATION

Although we have no reference frame on Saturn from which to measure the wind speed, we do know that the speeds are twice as large as those on Jupiter. This is a remarkable result, given that the planets are so similar and the power per unit area at Saturn is less than one-third that at Jupiter. On Saturn, features near the equator go around with periods as short as 10h 11m, and features near ±35° go around with periods as long as 10h 41m. This large difference in rotation periods implies large winds, but the value of the wind depends on the reference period. If the equatorial period were the reference, then there would be currents at ±35° flowing to the west at 396 m s^{-1}. If the ±35° period were the reference, then there would be a current at the equator flowing to the east at 481 m s^{-1}. The difference arises because one is defining the reference frame based on an angular velocity rather than a linear velocity. Either way, the winds on Saturn are larger than the winds on Jupiter. There, the maximum eastward jet speed is 163 m s^{-1}, and the maximum westward jet speed is 57 m s^{-1}. The velocity difference—the sum of the two speeds—is 220 m s^{-1}, which is one-half the velocity difference on Saturn.

Following the Voyager encounters in 1980–1981, the winds have been measured relative to a reference frame rotating with a period of 10h 39m 22.4s (fig. 7.6). This period was associated with radio emissions at kilometer wavelengths—the SKR (Saturn kilometric radiation), which was assumed to be rotating with the internal

219

magnetic field. However, during the fifteen years after the Voyager encounters it became clear that the SKR period was drifting, and Cassini revealed that there were two SKR periods,[108] one arising from radio sources in the northern hemisphere and the other from sources in the southern hemisphere. In 2007–2008 the northern period reached a minimum of ~10h 34m, and the southern period reached a maximum of ~10h 49m, at which point they started to converge. They crossed over in February 2010, when both sources had periods of ~10h 41.5m. This kind of variability could not have arisen from the main planetary dynamo, and is almost certainly a seasonal effect in the planet's magnetosphere and ionosphere. The Voyager SKR period is no longer regarded as the period of Saturn's interior, but it is still used as a reference frame for measuring positions and velocities.

If the SKR period were the period of Saturn's interior, then the atmosphere would be rotating faster than the interior at almost all latitudes. The winds would be strong and eastward relative to the internal "level of no motion." Saturn wouldn't look like Jupiter, which has a more balanced profile of alternating eastward and westward jets. The Voyager imaging team argued that the winds could not drop to zero in a thin layer immediately below the clouds, if zero meant co-rotating at the SKR period.[109] The argument is based on the thermal wind equation (box 8.6), which says that the increase of eastward wind with altitude is proportional to the decrease of temperature with latitude. The latter has been measured, at least at the cloud top levels, and it is small. Therefore the rate of increase of eastward wind with altitude must also be small. The strong eastward winds must persist to great depths, according to this argument. Using the observed upper bound on the temperature gradient, the Voyager imaging team concluded that the pressure at the level of no motion defined by the SKR period must be at least a thousand times the pressure at cloud base.[109]

One could reduce the required depth to the level of no motion by reducing the velocity u. This could be achieved by increasing the assumed rotation rate of the interior, that is, by abandoning the SKR period and choosing a shorter period. The Voyager imaging team considered this possibility as well. There is one period for which the average Coriolis force due to the westward jets exactly balances that due to the eastward jets— the integral of $2\Omega \sin\phi \, u$ with respect to latitude is zero. The imaging team estimated that an internal period between 10h 31m and 10h 32m would do this. Choosing an intermediate value for the reference period does not make the winds go away. The difference between the strongest eastward current and the strongest westward current is still greater than 400 m s^{-1}, but now the eastward and westward currents are more balanced.

The Voyager approach uses the winds at cloud top level to infer the interior rate of rotation. Another approach uses the measured shape of a constant pressure surface, for example, the 100-mbar surface.[110] The inferred interior period is that whose equilibrium shape most nearly matches that of the 100-mbar surface. A third approach requires that the measured atmospheric jets satisfy a particular stability criterion that depends on the wind speed.[III] All three approaches use measurements made at cloud-top levels to infer the internal rotation period. No approach uses information from levels below the 1 bar pressure level, because no useful information from those levels exists. However the connection between the cloud top levels and the planetary interior is tenuous. If one tried the same approach at Earth, using the height of the 500-mbar surface or the winds at that level, one would infer a rotation period of 23.2 hours. If one used the 100-mbar surface, one would infer a period of 22.8 hours. The problem is that the atmosphere at these levels rotates faster than the solid planet—the winds at these levels generally blow to the east. This could be happening on Saturn, although there is no solid planet and the level at which the gas rotates as a solid body is unknown. The internal period could be quite different from the atmospheric periods if the transition to solid body rotation is deep.

The internal rotation rate might not matter for the climate, especially if the level of no motion begins at great depth below cloud top. What matters more is the state of the atmosphere just below the clouds, where the rain is evaporating, and how the subcloud region interacts with the clouds that we see. Either way, the climate of Saturn is significantly different from that of Jupiter. Somewhere there are enough fundamental differences between the two planets to account for Saturn's high wind speeds, violent outbursts every twenty to thirty years, and lack of permanent features like Jupiter's Great Red Spot and white ovals.

10 URANUS, NEPTUNE, AND EXOPLANETS

10.1 URANUS

Uranus is the planet that spins on its side (fig. 10.1). The rotation axis of the planet is tipped at an angle of 98° with respect to the orbit axis. In other words, the obliquity is 98°. The only other planet with an obliquity greater than 90° is Venus, whose obliquity is 177°. Venus might have evolved gradually into its current spin state. It is close enough to the Sun that tides could have a big effect on its rotation. The extreme length of the day at Venus—117 Earth days—is evidence that the spin of Venus has evolved. In contrast, Uranus is only weakly affected by tides from the Sun, because it is so far away, and is more likely to have been knocked on its side during a giant impact. Models of planet accretion give a gradual clumping of small bodies into medium-sized bodies and then into large bodies, until finally only a few large bodies are left. The final collisions, which involved these large bodies, would have been quite violent and were capable of knocking Uranus on its side.

Once during the Uranian year—a period of eighty-four Earth years—the Sun is almost directly overhead at the north pole, and half a year later the Sun is almost

 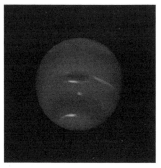

Figure 10.1. Uranus (*left*) and Neptune (*right*). The Uranus image was taken in 1998 by the Hubble Space Telescope. The Neptune image was taken in 1989 by Voyager. The winds of Neptune, measured by cloud tracking (fig. 7.6), are the fastest in the solar system.

(Sources: http://hubblesite.org/gallery/album/solar_system/ pr1998035a/, http://photojournal.jpl.nasa.gov/, PIA02245)

directly overhead at the south pole. If Saturn had a Tropic of Cancer or a Tropic of Capricorn, they would be at latitudes of ±82°. Averaged over the year, the high latitudes get more sunlight than the low latitudes, which is opposite to Earth and all the other planets. One might think this pattern of solar heating would have an enormous effect on the climate, but in fact, Uranus looks like a tipped-over version of Jupiter and Saturn (fig. 10.1). The clouds are arrayed in bands parallel to latitude circles, and the jets blow in the east-west direction, showing that rotation controls the weather and dominates solar heating,[112] at least at cloud-top levels.

What might happen when the Sun beats down on one hemisphere for forty-two Earth years? One might expect that the summer hemisphere would be the warmest place on the planet. But Uranus is 19.2 AU from the Sun, so sunlight is almost four hundred times fainter than it is at Earth. The depth over which sunlight is absorbed is uncertain, but 2 bars is a reasonable value. Averaged over the half-year that the pole is in sunlight, a layer of atmosphere extending down to the 2 bar level will heat up by only 5 K. There are two reasons this number is so small. First, the sunlight is fainter, and second, the hydrogen-helium atmosphere has a high specific heat. Per molecule, gases hold roughly the same amount of internal energy (box 2.1), but per mass the lighter gases hold more. The high specific heat limits the temperature change when a given amount of energy is absorbed.

In talking about the winds on Uranus, it is necessary to use the right notation, because Uranus, like Venus, spins backward as viewed from above the plane of the solar system. By definition, we are viewing the north pole, but the backward spin means that the direction of rotation is to the west, not to the east. Therefore, in comparing the winds on Uranus with the winds on Earth, we will use the words "prograde" and "retrograde" instead of "east" and "west." Prograde is always in the direction of rotation and retrograde is always the opposite.

Having the summer pole warmer than the equator is different from the Earth's troposphere, the part of the atmosphere within ~10 km of the surface. There, both poles are colder than the equator, even in summer. The thermal wind equation (box 8.6) implies that the prograde winds become stronger with height, which largely explains the midlatitude jet streams in each hemisphere. The thermal contrast is greater during the winter than during the summer, so the jet stream is strongest during the winter. But now consider our hypothetical Uranus, whose summer pole is warmer than the equator. Then the thermal wind equation implies that the retrograde winds become stronger with height. One might expect a retrograde midlatitude jet stream in the summer hemisphere. The winter hemisphere would be more like Earth, where the pole is colder than the equator and the midlatitude jet stream is prograde.

Every half-year, the poles switch positions relative to the Sun. This means that the warm pole with the retrograde jet becomes the cold pole with the prograde jet, and vice versa. The jet directions would reverse every year with the seasons. This reversing jet model does not agree with observation, either for Uranus or for Earth, at least not in the Earth's troposphere—the part of the atmosphere where we live and where commercial airplanes fly. It does work for the Earth's mesosphere—the layer above the stratosphere—where the pressure is in the range 0.1–1.0 mbar. The air in the mesosphere is warmest over the summer pole and coldest over the winter pole, and there is a retrograde jet in the summer hemisphere and a prograde jet in the winter hemisphere. The temperature gradient reverses with the seasons, and this causes a reversal of the mesospheric jets. As we know, the Earth's troposphere is always warmest over the equator, and this causes prograde jets in both hemispheres. The difference is that density in the Earth's mesosphere is so low that small amounts of absorbed sunlight can produce large temperature changes that follow the Sun. The pressure in the troposphere is a thousand times greater than in the mesosphere, and the equator is always warmer than the poles.

Although we have observed the zonal jets for less than half a Uranian year, it appears that the zonal jet profile is symmetric about the equator, and the jets do not reverse direction with the seasons.[112] There is apparently one prograde jet in each hemisphere, and it is at latitudes above 30° (fig. 7.6). In between, there is a broad retrograde jet at latitudes up to 30°. Despite its 98° obliquity, the zonal jet profile of Uranus resembles that of Earth more than it resembles those of Jupiter and Saturn. Surprisingly, Neptune, with its 30° obliquity, resembles Uranus and Earth, and not Jupiter and Saturn.

These simple models fail to explain the zonal jets of Uranus. They all assume that the large seasonal excursion of the Sun has a large effect on the circulation, but apparently it doesn't. The wind profile—zonal wind versus latitude—is symmetric about the equator and doesn't change much with the seasons. Uranus is like the Earth, even though the Earth gets more sunlight at the equator and has continents and oceans that break up the flow. One possibility is that the zonal winds on both planets are driven by eddies, and the process that converts eddy energy into zonal energy is the same whether the equator is warmer than the poles or vice versa. The eddies could get their energy by organizing the convection that brings heat up from below, or they could get their energy from the temperature difference between the poles and the equator. However, no temperature difference was seen at the time of the Voyager encounter. Uranus shows us that the connection between absorbed sunlight and the wind is subtle. Sunlight may be the ultimate energy source for the wind, but the rotation of the planet determines the wind pattern.

10.2 NEPTUNE

Neptune is the end of the line for giant planets in our solar system. It is farthest from the Sun and receives only ~5% of the power per unit area that Jupiter receives. The temperature at which it radiates to space is 59 K, which is slightly more than half the radiating temperature of Jupiter. Neptune's bulk density is higher than those of the other giant planets, indicating that it has a higher fraction of heavy elements—heavier than H and He. The abundance of methane, the only heavy gas that we can measure, is ~100 times the abundance in Jupiter's atmosphere.[79] Yet for climatologists, the most interesting feature of Neptune is the high wind speed (fig. 7.6). In fact, Neptune is probably the windiest planet in the solar system.[113] The difference between the fastest prograde jet and the fastest retrograde jet is ~600 m s^{-1}, which is more than ten times the difference on Earth and 50% greater than the difference at Saturn. Saturn's winds are uncertain because we don't know the rotation rate of the interior, but it appears that Neptune's winds are stronger.

Why should the wind get stronger as one moves to planets farther from the Sun? Why does Earth have the weakest winds in the solar system, despite receiving ~900 times more sunlight (power per unit area) than Neptune? The answer must involve the balance between forcing and dissipation. Forcing includes all the energy sources that warm the air in some places and cool it in others, leading to pressure gradients that accelerate the flow. Dissipation includes all the processes that reduce the kinetic energy of the air and turn it into heat. In a planetary atmosphere, the dissipation arises from turbulence, which is more effective than the viscosity in reducing kinetic energy. Somehow, and this is the puzzle, as one moves out in the solar system the dissipation decreases faster than the forcing. With less dissipation, the wind at Neptune reaches the highest speeds.

There is no direct way at present to test this idea, since the dissipation may be occurring at small scales, below the limit of resolution of cameras that are on the ground or in space. Also, the dissipation may be occurring below the tops of the clouds, either in the convection zone or deeper if waves are carrying the energy downward. We can probe the atmosphere with radio waves and detect turbulence from the fluctuations. And we can also just look at the atmospheres. Certainly Jupiter looks more active than Neptune. Usually just a handful of spots are visible on Neptune at any one time. On Jupiter the number is a thousand times greater. The spots are clouds, and they may be hundreds of km in diameter, but they could be related to the dissipation. So if a planet responds to reduced sunlight by reducing the dissipation, then the large-scale winds might get stronger. This might seem like a strange result, but the facts are strange: Jupiter has the weakest winds of the giant planets, and Neptune has the strongest.

Without turbulence, the zonal winds on a giant planet could coast along for hundreds or thousands of years. The molecular viscosity is low, and there are no continents for the winds to rub up against and no waves for the winds to blow against the shore. On a large scale, the thunderstorms, tornados, and hot air rising might look like dissipating mechanisms. With less sunlight, these mechanisms are weaker. The reduced dissipation at Neptune might outweigh the reduced forcing, and the winds might just be coasting along.

Neptune's spots give further support to the low-dissipation hypothesis. Before the age of computers, applied mathematicians solved fluid dynamics problems with pencil and paper. They couldn't handle turbulence, but they could solve problems of inviscid (zero-viscosity) flow. An example is an elliptical vortex in a zonal shear flow.[89] The ratio of the vorticity (spin of fluid elements) inside the vortex to the spin of the shear flow outside determines the ratio of the long axis of the ellipse to the short axis. If one perturbs the shape, it will expand and contract, rocking back and forth about a state where its long axis points east-west. Since the fluid has no viscosity, the oscillations continue indefinitely. Surprisingly, Neptune's Great Dark Spot was behaving this way during the Voyager encounter (fig. 10.2). Some of the spots on other planets oscillate, but no other spot mimics so accurately the inviscid spots in the pencil and paper theory.

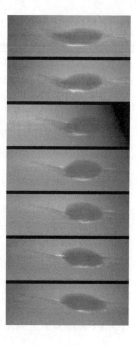

Figure 10.2. Changes of Neptune's Great Dark Spot over a 4.5-day interval. The long axis of the spot is ~10,000 km. The changes are part of a longer cycle in which the aspect ratio and orientation of the spot oscillate with a 7.5-day period. The oscillations are successfully described by a simple dynamical model that has no forcing or dissipation.[89]

(Source: http://photojournal.jpl.nasa.gov/, PIA00045)

This is a good point at which to review the weather and climates of the giant planets, and maybe to identify the controlling factors. The incident sunlight has surprisingly little effect. Contrary to intuition, the winds get stronger as you move away from the Sun: Neptune has the strongest winds of the giant planets, and Jupiter has the weakest. Furthermore, the dramatic seasons on Uranus seem to have little effect on the pattern of the winds. Uranus looks like a tipped-over version of the other giant planets. Rotation controls the pattern of the winds, forcing the flow into jets that circle the planet at constant latitude. The number of jets and the direction of the jet at the equator are different. Jupiter has five or six eastward jets in each hemisphere, Saturn has four, Uranus has one, and Neptune has one. The equatorial jets on Jupiter and Saturn are prograde, and those on Uranus and Neptune are retrograde. Neptune and Uranus have velocity profiles that look like the Earth's, even though the Earth has continents and the giant planets don't. The only obvious controlling factor is rotation. The obliquity has little or no effect, and the seasons have little or no effect. Either there are additional variables that are causing the differences between the giant planets, or else the differences are the result of chance. If that is the case, in a few thousand years, the climates of the giant planets may be different from what they are today.

10.3 EXTRASOLAR PLANETS

Exoplanet research is an exploding field. The number of known planets around other stars has more than doubled every two years since the first discoveries in the mid-1990s, and the exponential pace shows no sign of abating. Most of the hard facts learned are about the orbits and bulk properties—mass, radius, and distance from the parent star—but data about climate variables like temperatures, composition, and atmospheric circulation are coming in. For this book, the choices were: to give a thorough review that becomes obsolete in a few years, to skip the subject altogether, or to describe it briefly and show the reader that the subject exists and to give him or her some idea of the territory ahead. This last is the path we have taken.

The first discoveries were of giant planets orbiting close to their parent stars. These planets are the easiest to detect because they exert the greatest force on their stars and cause the greatest Doppler shift in the stars' spectral lines. This is the so-called radial velocity method of planet detection. It takes advantage of the fact that the star and planet orbit around their common center of mass, which is displaced from the center of mass of the star. Unless the orbit axis of the planet is pointing directly toward the Earth, we see the star move toward and away from us once per orbit. The light from the star is Doppler shifted toward shorter wavelengths when the star is approaching and is shifted toward longer wavelengths when the star is receding. Since the starlight has narrow spectral lines at definite wavelengths, the shift is detectable. Currently, one can detect stellar velocities as small as 1 m s^{-1}, which is the speed of a person walking.

The radial velocity method gives the period of the orbit and the magnitude of the Doppler shift. Interpreting these quantities requires knowing the mass of the star, and for main-sequence stars, the color gives a good estimate of the mass. From the star's mass and the period of the orbit one can compute the distance of the planet to the star, and from the Doppler shift, one can estimate the value of $M \sin i$, where M is the planet's mass and i is the angle between the planet's orbit axis and the line of sight to the Earth. Since $\sin i$ is unknown, the value of $M \sin i$ is a lower bound on M. Since the mid-1990s the radial velocity method has yielded hundreds of planets larger than Jupiter and closer to their stars than Mercury. In many cases the Doppler shifts are more complicated than simple sine waves, indicating that the planets are in eccentric orbits or in multiplanet systems. Some of the planets orbit their stars with periods of a few days. The intense stellar light heats these planets to temperatures well above 1000 K, and they are appropriately called hot Jupiters.

The radial velocity method relies entirely on light from the star; photons from the planet remain undetected. To learn more about the planet, the observers began using the transit method, which relies on those special planets that have $\sin i \approx 1.0$ and therefore pass in front of their stars as viewed from Earth. A typical transit lasts for a few hours, so ideally, one monitors a large number of stars every few minutes and searches for dips in brightness as the planet blocks some of the starlight headed our way. Currently, a dip as small as one part in ten thousand is detectable. Figure 10.3 shows Venus transiting in front of the Sun in 2012. We cannot image stars as we can the Sun, so all we can see is the dip in the starlight (fig. 10.4). The magnitude of the dip measures the area subtended by the planet compared to that of the star. Moreover, the radial velocity method for a transiting planet gives M directly, since $\sin i \approx 1.0$. Thus we can get the density of the planet.

If we watch the transit at different wavelengths, we see subtle differences in the magnitude of the dip. These can be attributed to absorption of starlight in the planet's

Figure 10.3. Transit of Venus observed from the International Space Station on June 7, 2012.

(Source: http://earthobservatory.nasa.gov/IOTD/view.php?id= 78196)

upper atmosphere: At wavelengths where the upper atmosphere is transparent, the starlight passes through. At other wavelengths the starlight is absorbed. Since each substance absorbs at specific wavelengths, in principle one can measure the composition of the planet's atmosphere. The limitation is that only the atmosphere of the planet is involved in the absorption. The bulk of the planet is opaque to the starlight, so the size of the absorption is proportional to the projected area of the atmosphere, which is small compared to the projected area of the planet, and that is small compared to the projected area of the star. Thus the signal-to-noise ratio is extremely small for this type of measurement.

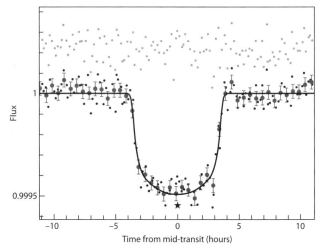

Figure 10.4. Light curve of Kepler-22b. From the period of the orbit and the depth of the light curve, one infers that Kepler-22b is a super Earth-sized planet in its star's habitable zone. Its equilibrium temperature is 262 K; its radius is 2.38 Earth radii, and its orbit period is 290 days. The small black dots are the observations; the large black dots are 30-minute averages, and the gray points are scaled-up residuals between the observations and the orbital model.

(Source: http://kepler.nasa.gov/Mission/discoveries/kepler22b/)

Fortunately, for every transit there is an eclipse, where the planet passes behind the star. Sometimes this event is called the secondary eclipse and the transit is called the primary eclipse. During the secondary eclipse, the light from the planet is blocked, so we can determine the brightness of the planet by subtraction. Brightness as a function of wavelength gives the spectrum. We get information about the atmospheric composition, the albedo, and the temperature. The planet is not as bright as the star, and the dip during the secondary eclipse is typically less than that during the primary eclipse, so the signal to noise is small. Still, the depth of the secondary eclipse is proportional to the projected area of the entire planet and not just its upper atmosphere, so the opportunity to do spectroscopy is generally better. Water vapor, methane, CO, CO_2, and atomic sodium have been detected in this way.

Like the radial velocity method, the transit method is biased toward hot Jupiters. Large planets block a larger fraction of the starlight, so the signal to noise is higher. Planets orbiting close to the their stars are more likely to pass in front of them—the constraint $\sin i \approx 90°$ is not as stringent when the planet is close. Also, it is easier to confirm that a dip is real when it repeats every few days. An extraterrestrial observer would have to wait a few years to confirm that Earth had passed in front of the Sun. Such confirmation is necessary because detectors can drift and the stars can be noisy—sometimes they are double stars, and often they have starspots. The hot Jupiter bias can be partly overcome because the technology has improved to the point where one can simultaneously monitor 150,000 stars every minute. The Kepler spacecraft was designed to do just that, monitoring a dense star field from above Earth's atmosphere where interference is minimal. In its first year of operation, Kepler released the locations of over two thousand candidate planets that appeared to be transiting their stars. Confirmation is required for planets farther from their stars, and the radial velocity method is required to determine mass and density, but Kepler has vastly expanded the field.

....

There are other methods for detecting exoplanets, like gravitational lensing and direct imaging. These methods are capable of detecting planets that are farther from their stars than the hot Jupiters. Kepler has detected Neptune-sized planets and a few super-Earths—planets less than twice the size of the Earth. At least one Earth-sized planet is far enough away from its parent star to be in the habitable zone, defined to be where temperatures allow liquid water on the surface.

This flood of new data has started a flood of new questions: What kinds of stars have planets? What fraction of stars have planets? How common are Earth-sized planets? Do extrasolar planets have moons? Do double stars have planets? How do planets form and how do they migrate into their current orbits? Do planets get ejected from their stellar systems? Do they get absorbed onto their parent stars? What fraction of planets are in the habitable zones? Do planets show signs of life? What would constitute a sign of life—oxygen atmosphere, disequilibrium species, signs of photosynthesis, reflection off an ocean? How would we know? This is not an exhaustive list. Some of the questions are being addressed now. Others are on hold, waiting for technological breakthroughs.

Taking all the extrasolar planets that have been detected, and allowing for the hot Jupiter bias, it appears that roughly half of the stars in the galaxy have planets, and half of the planets are super-Earths or smaller. It is likely that many are in the habitable zones of their parent stars. The number and diversity of exoplanets is much greater than the number and diversity of planets in our solar system. The diversity of planetary climates is undoubtedly just as great. We study climates in this larger context to learn how climates work and also to know whether climates like ours are common throughout the universe. If they are, then extraterrestrial life may be common as well.

11 CONCLUSION

THE FIRST BIG LESSON FROM EXPLORING THE PLANETS IS humility: No matter how much you know from studying the Earth, you will be surprised. Your predictions about what you will find will often be wrong since those predictions are extrapolations of processes and environments found on Earth. Since the Earth doesn't exhibit the same range of possibilities that the planets do, the predictions require both knowledge and imagination. Either our knowledge of Earth science has been inadequate or our ability to extrapolate it to other planets has been inadequate because our collective imagination did not prepare us for what we found.

To be fair, we usually got the general principles right and failed more on the specifics. One general principle is that size matters. The larger objects are able to hold on to their atmospheres better than the small objects. The atmospheres of small objects evolve as their lighter atmospheric constituents escape into space. Another principle that held: Distance from the Sun matters. The outer solar system is hydrogen rich and the inner solar system is oxygen rich; as one moves away from the Sun different substances take on different roles, going from being primarily gases to being primarily liquids, and then to being primarily solids. A third correct principle: Bad stuff happens. Feedback happens, and tipping points exist. The distribution of planetary climates is not smooth. There are boundaries in the controlling parameters, with radically different climates on either side. We knew these general principles before the age of space exploration, but we let our terrestrial biases affect their application.

Often we assumed the planets were more Earth-like than they turned out to be. Venus was assumed to have a humid atmosphere, liquid water, and warm temperatures beneath its clouds, which were supposed to be made of condensed water. This benign picture came apart in the 1960s when radio telescopes peering through the clouds measured brightness temperatures close to 700 K. Also in the 1960s, the way sunlight scattered off the clouds revealed that they were made of sulfuric acid. The Soviet Venera probes showed that the atmosphere was a massive reservoir of carbon dioxide, exceeding the reservoir of limestone rocks on Earth. The U.S. Pioneer Venus radar images showed a lightly cratered volcanic landscape with no trace of plate tectonics. In all these cases, the observations drove the theories, which came later.

The expectations for Mars were even higher than those for Venus. Seasonal changes on the surface were interpreted by some as irrigation canals built by intelligent beings. To their credit, most planetary astronomers disputed these interpretations, and controversies arose over whether water vapor had been detected in the Martian atmosphere or not. The U.S. Mariner 4 spacecraft flew by Mars in 1965 and revealed a heavily cratered, geologically dead surface. From the delay of the spacecraft's radio signal passing through the atmosphere, it became clear that the surface pressure was ~6 mbar, more than an order of magnitude less than the previously accepted value.[114] It also became clear that carbon dioxide was the main constituent of the atmosphere, and that it would condense out at a temperature of 148 K. With permanent dry ice deposits at this temperature, any water vapor in the atmosphere would be cold-trapped out.[30] Later spacecraft softened this view. Not all the surface was geologically dead, and fluvial features were found. The optimism persisted, and the Viking landers were equipped with instruments designed to detect Martian life. Viking was a scientific success, but no life was detected. As we have seen, liquid water on the surface of Mars is still a possibility, and evidence for a warm, wet early Mars continues to build. One of the attractive theories from the 1960s, that a massive reservoir of CO_2 at the poles could be controlling the mass of the CO_2 atmosphere,[30] had to be shelved because the massive reservoir couldn't be found. The search goes on for buried CO_2 ice and buried carbonate rocks, so the issue is not completely settled.

The amateur astronomers had been following the spots and bands on Jupiter for a hundred years before the era of large telescopes and interplanetary spacecraft. The amateurs were cautious about calling them weather features, and used more descriptive terms like spots, belts, zones, plumes, festoons, barges, and streamers. Nevertheless, by the 1960s it was clear that we were seeing the atmosphere in motion. The long life of the weather features was so unlike that of weather features on Earth that people invoked steering by the underlying topography of the solid planet.[115] However, by the mid-1960s it had become clear that there was no solid planet. There was no way to match the planet's low density except with hydrogen and helium, and neither of these would form solids under Jovian conditions. The Pioneer and Voyager encounters in the 1970s deepened the mystery by showing that the atmosphere was turbulent—full of waves and eddies that could have destroyed the flow features like the zonal jets and the Great Red Spot. Here the theorists had some success. Numerical models showed that the eddies could be pumping energy into the zonal jets, not the other way around.[91,92] And the models showed that stable vortices like the Great Red Spot could arise spontaneously from unstable zonal jets.[86,87,88] Now the question is whether the flows are shallow or deep—whether the high winds are confined to the weather layer 100 km thick where clouds are forming, or whether they extend many thousands of km into the fluid interior.

Our expectations for the moons of the giant planets were more pessimistic than our expectations for Mars and Venus. We thought most of the moons were battered ice balls whose cold surfaces hadn't changed since the formation of the solar system. The moons were assumed to be like our own moon, and were not supposed to be Earth-like. The theorists did anticipate that Io would be dissipating a large amount of tidal energy in its interior,[116] and this was confirmed by Voyager's discovery of active volcanism in 1979. The idea of the habitable zone as a place between Mars and Venus where water could exist as a liquid slowly gave way in the 1990s. From the gravity and magnetic fields of the large satellites it became clear that they have liquid water beneath their icy crusts.[117] How deep the water lies is still uncertain, but ~10 km is a possibility. We now think of the habitable zone as an archipelago that includes the Earth but also includes the icy moons with their liquid water oceans.

The search for extrasolar planets took off in the mid-1990s. The first discovery of a planet orbiting a main sequence star was in 1995. It is not clear whether the explosion of discoveries could have happened decades earlier, or whether the observers had to wait for developments in technology. Previous searches had focused on stellar systems like our own, with Jupiter-sized planets orbiting at distances of 10 AU or greater and maybe a few Earth-sized planets closer in. Such objects are almost undetectable even today, because their pull on the parent star is so weak and their orbits are so long. One has to wait a professional lifetime to observe a complete orbit of Saturn, for instance. The breakthrough came because there were many giant planets close to their parent stars and because these planets were much easier to detect. To a certain extent, no one was looking in that region of parameter space until the first discovery.

Does the study of planetary climates provide lessons for Earth? Yes, if one takes a broad view. The planets show us what atmospheres are capable of. Winds of 100 m s^{-1} are common throughout the solar system, and the outer planets have winds in the 400–500 m s^{-1} range. Could such winds occur on Earth? Probably not, but it would be nice to know why. Atmospheres are capable of evolving in horrendous ways. A runaway greenhouse apparently occurred on Venus. Could it occur on Earth? Again, probably not, but one should think about it. On the brighter side, on some planets the weather patterns persist for decades or centuries, and are highly predictable. On Earth one talks about the theoretical limit to predictability and assumes it is a few weeks. Do we understand where that limit comes from? Can we extend the limit to match the giant planets? Probably not, but it's worth investigating.

Has studying planets informed us about the likelihood of extraterrestrial life? The answer is a resounding yes. Knowing that extrasolar planets are abundant is a huge positive step. There are billions of galaxies and billions of stars in each galaxy. Now it appears that there are billions of planets in each galaxy as well. This changes the odds dramatically.

Has studying the planets in our solar system made the development of extraterrestrial life seem more likely? Yes again. We have discovered that the basic elements of life, C, H, O, and N are common throughout the solar system. These are also the basic elements of climate, and they appear in molecular form as CH_4, CO_2, H_2O, NH_3, N_2, and O_2. We have discovered many places in our solar system where water exists or existed as a liquid, although many of those places have dried up or gone underground. Underground life needs chemical energy, since there is no sunlight, and that could be a problem. Nevertheless, life evolved on Earth soon after the formation of the solar system, so with the right raw materials, life seems almost inevitable. Turning the argument around, we have not found any showstoppers. There is no evidence that terrestrial life required improbably special circumstances. In fact, Earth seems to be a typical planet orbiting a typical star, with a typical collection of raw materials that constitute the building blocks of life. There is no reason to believe we are alone.

adiabatic lapse rate—the decrease of temperature with altitude for a fluid element that is not exchanging heat with its surroundings.

albedo—the fraction of incoming light that is scattered and not absorbed; used mostly for short wavelengths, i.e., sunlight.

Astronomical Unit (AU)—the mean distance between the Earth and the Sun; more precisely it is the semi-major axis of the Earth's orbit.

beta effect—fluid dynamical effect of the planetary vorticity increasing with latitude; causes the relative vorticity of fluid elements to change as they change latitude.

blackbody radiation—radiation emitted by a perfect absorber; also, the equilibrium radiation inside an isothermal cavity.

boiling—formation of vapor bubbles within a liquid; requires that the saturation vapor pressure be equal to or greater than the total ambient pressure.

Boltzmann's constant—denoted by the symbol k_B. It is the constant entering in the ideal gas equation of state and the Maxwell-Boltzmann distribution of velocities, such that the average kinetic energy per molecule is $(3/2)k_B T$.

buoyancy—net vertical force on a fluid element in a gravitational field when its density is different from that of its surroundings.

centrifugal force—apparent force acting on all bodies in a rotating system. Compare with the Coriolis force, which acts only on moving bodies in the rotating system.

Clausius Clapeyron equation—thermodynamic relation giving the change of saturation vapor pressure with temperature.

convection—circulation of fluid elements driven by buoyancy forces that arise from temperature differences within the fluid.

Coriolis force—apparent force acting on moving bodies in a rotating system. Compare with the centrifugal force, which acts on all bodies.

cyclonic vorticity—spin about a vertical axis that is in the same direction as the vertical component of the planet's spin; anticyclonic vorticity is when the two spins are opposite.

deuterium—isotope of hydrogen whose nucleus consists of one proton and one neutron; has approximately twice the mass of an ordinary hydrogen atom.

Doppler shift—change in the frequency of a wave when the source is moving with respect to the receiver.

eddy—departure of fluid motion from a uniform mean state. In atmospheric science, the mean state is usually defined as a steady zonal flow, i.e., a time-independent flow in the east-west direction.

enrichment factor—term used to describe an element's abundance relative to some standard. Often the standard is the element's abundance on the Sun.

equation of state—relation between pressure, density, temperature, and other thermodynamic variables in equilibrium. For a pure substance, only two variables are independent; the equation of state gives the values of the others when two are known.

escape velocity—velocity required for an object to escape the gravitational field of a planet.

exosphere—the outermost region of an atmosphere where molecules do not collide with each other; they emerge from the layers below and follow ballistic trajectories before falling back into the denser part of the atmosphere.

faint young Sun paradox—disagreement between geologic evidence of warm climates of the distant past and astronomical evidence that the Sun's luminosity was only 70% of its current value during the same period.

feedback—causative loop in which the output of a process modifies the input. In a positive feedback the output strengthens the input; in a negative feedback the output weakens the input.

flux of radiation—radiant energy crossing unit area per unit time. Units are power per unit area. Flux is also called irradiance. It is an integral over separate beams each propagating in its own direction.

fractionation factor—in isotope geochemistry, it is the ratio by which the relative abundance of two isotopes differs between two phases of a system. For example, the hydrogen and oxygen isotope ratios differ between the vapor phase and the liquid phase when the two phases are in equilibrium with each other.

gas constant—a constant equal to pressure divided by the product of density and temperature. It is a different constant for each gas, and is equal to Boltzmann's constant divided by the mass of the molecule.

geostrophic balance—balance between the horizontal component of the Coriolis force and the horizontal pressure gradient force. Geostrophic balance is a useful approximation for large scales and slowly varying processes away from the equator.

greenhouse gas—a gas that absorbs infrared light and prevents radiation emitted deep in the atmosphere from escaping to space.

habitable zone—region around a star where temperatures allow water to exist primarily as a liquid. It is a function of distance to the star and the star's luminosity.

Hadley circulation—a large-scale overturning of the tropical atmosphere, with rising motion near the equator and sinking motion in the subtropics.

hydrostatic balance—a balance between the weight of the air and the vertical pressure-gradient force. Hydrostatic balance is a good approximation on large scales.

infrared—wavelengths longer than those of the visible range.

intensity of radiation—the flux associated with an individual beam, per unit solid angle of the beam. Units are power per unit area per steradian. Intensity is also called radiance or brightness.

isotopes—atoms of the same element with different numbers of neutrons in the nucleus. The chemistry is the same, but the masses of the atoms are different.

jet stream—a strong riverlike current of air. On most planets the jet streams blow primarily in the east-west direction.

Kelvin temperature scale—temperature measured from absolute zero. Units are defined to be the same as degrees Celsius, which is accomplished by having $T = 273.16$ K at the triple point of water.

lapse rate—rate of temperature decrease with altitude.

latent heat—energy required to vaporize a unit amount of condensate without changing the temperature or pressure. Units are energy per unit mass. The same energy is released when the vapor condenses.

luminosity—total power output of a star.

millibar—unit of pressure, equal to 10^{-3} bars or 10^2 Pa.

oblateness—dimensionless measure of the equatorial bulge. Specifically it is the difference between the equatorial and polar radii divided by the equatorial radius.

obliquity—angle between the spin axis of a planet and the axis of the orbit.

optical depth—dimensionless measure of the opacity of a medium. If there are no additions to the beam, its intensity decreases as e ($2.718\ldots$) to the power minus the optical depth as it passes through the medium. Optical depth is an integral of the absorption coefficient and the density of the medium along the path of the beam.

parcel—pointlike mass of fluid, also called a fluid element. A parcel contains a sufficient number of molecules to have a well-defined velocity, temperature, pressure, and other thermodynamic variables.

partial pressure—contribution of one gas to the total pressure of a mixture.

photodissociation—breaking of molecular bonds by absorption of radiation. The molecule usually splits into two pieces that fly apart at high speed.

Planck function—intensity of blackbody radiation; it is a function of temperature and wavelength (or frequency) only, and is a useful approximation to actual radiation from planetary surfaces and atmospheres.

planetary vorticity—the spin (vorticity) of a fluid element arising from its being on a rotating planet. It is denoted by the symbol f. The other part of its spin, known as the relative vorticity ζ, arises from its rotation relative to the planet.

potential vorticity—absolute vorticity (sum of planetary and relative vorticities) of a fluid element divided by the product of density times thickness. Potential vorticity (PV) is a conserved quantity, meaning that the fluid element retains its value of PV until it mixes with other elements or is modified by viscosity or heat transfer.

precipitable water—amount of water vapor in the atmosphere expressed as the depth of the liquid layer that would result if all the vapor were to condense out.

prograde—in the same direction as the planet's rotation.

radiative transfer—heat transfer by electromagnetic radiation. The theory of radiative transfer focuses on the flow of energy and ignores the wavelike nature of light.

Rayleigh-Kuo stability criterion—a necessary condition for a zonal flow to be unstable. It is not a sufficient condition for instability. Some flows violate the Rayleigh-Kuo stability criterion and are nevertheless stable.

relative vorticity—vorticity of a fluid element due to its spin relative to the rotating planet. The other part of the vorticity is that due to the spin of the planet itself. See planetary vorticity and potential vorticity.

retrograde—in the direction opposite to the planet's rotation. A retrograde wind is not a reversal of the rotational motion; the fluid still rotates in the direction of rotation, but it rotates more slowly than the planet.

Rossby number—inverse measure of the importance of rotation to the dynamics of the fluid. It is essentially the ratio of two time scales—the period of the planet's rotation (the day) divided by the time scale of the fluid motion. When the Rossby number is small (less than one), the effect of rotation is large. It is usually expressed as the typical velocity divided by the typical length scale of the flow, divided by the planetary vorticity f.

saturation vapor pressure (SVP)—the pressure of a vapor when it is in equilibrium with a condensed phase, either solid or liquid. In a mixture of gases, it is the partial pressure of the vapor in equilibrium with its own condensed phase.

scale height—the e-folding thickness of an atmosphere. The natural logarithm of pressure falls off with altitude as one over the scale height.

solar composition—chemical composition resulting from cooling a piece of the Sun down to planetary temperatures.

solid angle—an angular patch of the sky or bundle of directions measured in square radians; also called steradians and abbreviated as sr.

specific heat—energy required to raise a unit mass of matter by one degree. The specific heat is greater when pressure is held constant than when volume is held constant, because in the former case, some of the energy goes into doing work on the environment and is not available for raising the temperature.

spectrum—result of decomposition of light into its individual wavelengths.

stability—tendency to return to the original state after a small perturbation. Instability is the opposite; the small perturbation will grow to finite amplitude.

sublimation—evaporation directly from the solid phase to the vapor phase with no intermediate melting step.

superrotation—rotation of an atmosphere that is faster than the rotation of the solid planet.

thermal wind—in a rotating system, it is the rate of change of horizontal wind with height and is related to the temperature gradient in a direction perpendicular to the wind.

tide—effect of one celestial body on another, either due to the pull of gravity and inertial acceleration or to the heating effect of the Sun or star. The tidal response is fixed relative to the line between the two bodies, and it will appear to propagate relative to the body on which the tide is occurring if it is rotating.

transit—one celestial body passing in front of another; technically a transit is the same as an eclipse, but transit is used more often when the smaller body passes in front of the larger one.

turbulence—motion on time and space scales that are small compared to the scales of interest. Hence, turbulence is that part of the motion that is treated statistically.

upgradient momentum transfer—transport of momentum from a region of lower momentum to a region of higher momentum; it tends to strengthen the difference in speed between adjacent jets.

viscosity—the friction force between layers of fluid sliding past each other.

volatile—tendency to fly away; having a high vapor pressure at low temperatures.

vorticity—a pointlike property describing the spin of a fluid element. Vorticity is twice the angular velocity when the fluid is spinning like a solid body.

weighting function—the fraction that each thin slab contributes to the outgoing radiation. It is a function of distance along the ray path and goes to zero deep within an absorbing medium.

zonal wind—wind in the east-west direction, as distinct from the meridional wind, which is in the north-south direction. A zonal average is an average in the east-west direction, i.e., an average with respect to longitude.

Notes

....

1. de Pater, I., & Lissauer, J. J. (2010). *Planetary Sciences, Second Edition*. Cambridge University Press, Cambridge, 647 pp.

2. Beatty, J. K., Peterson, C. C., & Chaikin, A., eds. (1999). *The New Solar System, Fourth Edition*. Cambridge University Press, 421 pp.

3. Kasting, J. (2010). *How to Find a Habitable Planet*. Princeton University Press, Princeton, 326 pp.

4. Sánchez-Lavega, A. (2011). *An Introduction to Planetary Atmospheres*. Taylor and Francis, Boca Raton, FL, 587 pp.

5. Bougher, S. W., Hunten, D. M., & Phillips, R. J., eds. (1997). *Venus II*. University of Arizona Press, Tucson, 1362 pp.

6. Kieffer, H. H., Jakosky, B. M., Snyder, C. W., & Matthews, M. S., eds. (1992). *Mars*. University of Arizona Press, Tucson, 1498 pp.

7. Baganal, F., Dowling, T. E., & McKinnon, W. B., eds. (2004). *Jupiter: The Planet, Satellites, and Magnetosphere*. Cambridge University Press, Cambridge, 719 pp.

8. Dougherty, M. K., Esposito, L. W., & Krimigis, S. M., eds. (2009). *Saturn from Cassini-Huygens*. Springer Dordrecht, Heidelberg, 805 pp.

9. Brown, R., Lebreton, J.-P., & Waite, J. H., eds.(2009). *Titan from Cassini-Huygens*. Springer Dordrecht, Heidelberg, 535 pp.

10. Randall, D., (2012). *Atmosphere, Clouds, and Climate*. Princeton University Press, Princeton, 277 pp.

11. Salby, M. L. (1996). *Fundamentals of Atmospheric Physics*. Academic Press, San Diego, 627 pp.

....

12. Holton, J. R. (2004). *An Introduction to Dynamic Meteorology, Fourth Edition*. Elsevier Academic Press, Amsterdam, 535 pp.

13. Vallis, G. K. (2006). *Atmospheric and Oceanic Fluid Dynamics*. Cambridge University Press, Cambridge, 745 pp.

14. Clark, B. G., & Kuzmin, A. D. (1965). Measurement of Polarization and Brightness Distribution of Venus at 10.6-cm Wavelength. *Astrophysical Journal*, 142(1), 23–44.

15. Hansen, J. E., & Hovenier, J. W. (1974). Interpretation of Polarization of Venus. *Journal of the Atmospheric Sciences*, 31(4), 1137–1160.

16. Donohue, T. M., & Pollack, J. B.. (1983). Origin and Evolution of the Atmosphere of Venus. In *Venus*, eds. D. M. Hunten, L. Colin, T. M. Donohue, & V. I. Moroz. University of Arizona Press, Tucson, pp. 1003–1036.

17. Owen, T. (1992). The Composition and Early History of the Atmosphere of Mars. In *Mars*, eds. H. H. Kieffer, B. M. Jakosky, C. W. Snyder, and M. S. Matthews. University of Arizona Press, Tucson, pp. 818–834.

18. Bockelee-Morvan, D., Gautier, D., Lis, D. C., Young, K., Keene, J., Phillips, … Wooten, A. (1998). Deuterated Water in Comet C/1996 B2 (Hyakutake) and Its Implications for the Origin of Comets. *Icarus*, 133(1), 147–162.

19. Donohue, T. M., Grinspoon, D. H., Hartle, R. E., and Hodges, R. R. (1997). Ion/Neutral Escape of Hydrogen and Deuterium: Evolution of Water. In *Venus II*, eds. S. W. Bougher, D. M. Hunten, L. Colin, T. M. Donohue, & R. J. Phillips. University of Arizona Press, Tucson, pp. 385–414.

20. Bertaux, J. L., Vandaele, A. C., Korablev, O., Villard, E., Fedorova, A., Fussen, D., . . . Rodin, A. (2007). A Warm Layer in Venus' Cryosphere and High-Altitude Measurements of HF, HCl, H_2O and HDO. *Nature*, 450(7170), 646–649. doi: 10.1038/Nature05974.

....

21. Ingersoll, A. P. (1969). The Runaway Greenhouse: A History of Water on Venus. *Journal of the Atmospheric Sciences*, 26(6), 1191–1198.

22. Kasting, J. F. (1988). Runaway and Moist Greenhouse Atmospheres and the Evolution of Earth and Venus. *Icarus*, 74(3), 472–494.

23. Seiff, A. (1983). Thermal Structure of the Atmosphere of Venus. In *Venus*, ed. D. M. Hunten, L. Colin, T. M. Donohue, and V. I. Moroz. University of Arizona Press, Tucson, pp. 215–279.

24. Gierasch, P., et al. (1997). The General Circulation of the Venus Atmosphere: An Assessment. In *Venus II*, ed. S. W. Bougher, D. M. Hunten, L. Colin, T. M. Donohue, and R. J. Phillips. University of Arizona Press, Tucson, pp. 459–500.

25. Schubert, G. (1983). General Circulation and Dynamical State of the Venus Atmosphere. In *Venus*, ed. D. M. Hunten, L. Colin, T. M. Donohue, and V. I. Moroz. University of Arizona Press, Tucson, pp. 681–765.

26. Taylor, F. W., Hunten, D. M., and Ksanfomaliti, L. V. (1983). The Thermal Energy Balance of the Middle and Upper Atmosphere of Venus. In *Venus*, ed. D. M. Hunten, L. Colin, T. M. Donohue, and V. I. Moroz. University of Arizona Press, Tucson, pp. 650–680.

27. Gierasch, P. J. (1975). Meridional Circulation and Maintenance of Venus Atmospheric Rotation. *Journal of the Atmospheric Sciences*, 32(6), 1038–1044.

28. Hartmann, W. K. (2005). *Moons and Planets, Fifth Edition*. Brooks/Cole, Thompson Learning Center, Belmont, CA, 428 pp.

29. Bahcall, J. N., Huebner, W. F., Lubow, S. H., Parker, P. D., & Ulrich, R. K. (1982). Standard Solar Models and the Uncertainties in Predicted Capture Rates of Solar Neutrinos. *Reviews of Modern Physics*, 54(3), 767–800.

30. Leighton, R. B., & Murray, B. C. (1966). Behavior of Carbon Dioxide and Other Volatiles on Mars. *Science*, 153(3732), 136–144.

31. Pollack, J. B., Kasting, J. F., Richardson, S. M., & Poliakoff, K. (1987). The Case for a Wet, Warm Climate on Early Mars. *Icarus*, 71(2), 203–224.

32. Tian, F., Claire, M. W., Haqq-Misra, J. D., Smith, M., Crisp, D. C., Catling, D., Zahnle, K., & Kasting, J. F. (2010). Photochemical and Climate Consequences of Sulfur Outgassing on Early Mars. *Earth & Planet. Sci. Lett.* 295, 412–418.

33. Laskar, J., Levrard, B., & Mustard, J. F. (2002). Orbital Forcing of the Martian Polar Layered Deposits. *Nature*, 419(6905), 375–377. doi: 10.1038/Nature01066.

34. Phillips, R. J., Davis, B. J., Tanaka, K. L., Byrne, S., Mellon, M. T., Putzig, N. E., . . . Seu, R. (2011). Massive CO_2 Ice Deposits Sequestered in the South Polar Layered Deposits of Mars. *Science*, 332(6031), 838–841. doi: 10.1126/Science.1203091

35. Ehlmann, B. L., Mustard, J. F., Murchie, S. L., Poulet, F., Bishop, J. L., Brown, A. J., . . . Wray, J. J. (2008). Orbital Identification of Carbonate-Bearing Rocks on Mars. *Science*, 322(5909), 1828–1832. doi: 10.1126/Science.1164759.

36. Owen, T., Biemann, K., Rushneck, D. R., Biller, J. E., Howarth, D. W., & Lafleur, A. L. (1977). The Composition of the Atmosphere at the Surface of Mars. *Journal of Geophysical Research*, 82(28), 4635–4639.

37. Boynton, W. V., Feldman, W. C., Squyres, S. W., Prettyman, T. H., Bruckner, J., Evans, L. G., . . . Shinohara, C. (2002). Distribution of Hydrogen in the Near Surface of Mars: Evidence for Subsurface Ice Deposits. *Science*, 297(5578), 81–85. doi: 10.1126/Science.1073722.

38. Smith, D. E., Zuber, M. T., Frey, H. V., Garvin, J. B., Head, J. W., Muhleman, D. O., . . . Sun, X. L. (2001). Mars Orbiter Laser Altimeter: Experiment Summary After the First

Year of Global Mapping of Mars. *Journal of Geophysical Research-Planets*, 106(E10), 23689–23722.

39. Paige, D. A., & Ingersoll, A. P. (1985). Annual Heat-Balance of Martian Polar Caps from Viking Observations. *Science*, 228(4704), 1160–1168.

40. Thomas, P. C., James, P. B., Calvin, W. M., Haberle, R., & Malin, M. C. (2009). Residual south Polar Cap of Mars: Stratigraphy, History, and Implications of Recent Changes. *Icarus*, 203(2), 352–375. doi: 10.1016/J.Icarus.2009.05.014.

41. Bibring, J. P., Langevin, Y., Gendrin, A., Gondet, B., Poulet, F., Berthe, M., . . . Team, O. (2005). Mars Surface Diversity as Revealed by the OMEGA/Mars Express Observations. *Science*, 307(5715), 1576–1581. doi: 10.1126/Science.1108806.

42. Byrne, S., & Ingersoll, A. P. (2003). A sublimation Model for Martian South Polar Ice Features. *Science*, 299(5609), 1051–1053.

43. Tillman, J. E., Johnson, N. C., Guttorp, P., & Percival, D. B. (1993). The Martian Annual Atmospheric Pressure Cycle: Years Without Great Dust Storms. *Journal of Geophysical Research-Planets*, 98(E6), 10, 963–10,971.

44. Smith, M. D. (2002). The Annual Cycle of Water Vapor on Mars as Observed by the Thermal Emission Spectrometer. *Journal of Geophysical Research-Planets*, 107(E11). Artn 5115 doi: 10.1029/2001je001522.

45. Ingersoll, A. P. (1970). Mars: Occurrence of Liquid Water. *Science*, 168(3934), 972–973.

46. Malin, M. C., & Edgett, K. S. (2000). Evidence for Recent Groundwater Seepage and Surface Runoff on Mars. *Science*, 288(5475), 2330–2335.

47. Zurek, R. W., Barnes, J. R., Haberle, R. M., Pollack, J. B., Tillman, J. E., & Leovy, C. B. (1992). Dynamics of the Atmosphere of Mars. In *Mars*, eds. H. H. Kieffer, B. M. Jakosky, C. W. Snyder, and M. S. Matthews. University of Arizona Press, Tucson, pp. 835–933.

48. Barnes, J. R. (1981). Midlatitude Disturbances in the Martian Atmosphere—A 2nd Mars Year. *Journal of the Atmospheric Sciences*, 38(2), 225–234.

49. Pankine, A. A., & Ingersoll, A. P. (2004). Interannual Variability of Mars Global Dust storms: An Example of Self-Organized Criticality? *Icarus* 170, 514–518.

50. Lunine, J. I., Stevenson, D. J., & Yung, Y. L. (1983). Ethane Ocean on Titan. *Science*, 222(4629), 1229–1230.

51. Hayes, A., Aharonson, O., Callahan, P., Elachi, C., Gim, Y., Kirk, R., . . . Wall, S. (2008). Hydrocarbon Lakes on Titan: Distribution and Interaction with a Porous Regolith. *Geophysical Research Letters*, 35(9). Artn L09204 doi: 10.1029/2008gl033409.

52. Wilson, E. H., & Atreya, S. K. (2009). Titan's Carbon Budget and the Case of the Missing Ethane. *Journal of Physical Chemistry A*, 113(42), 11221–11226. doi: 10.1021/Jp905535a.

53. Porco, C. C., Baker, E., Barbara, J., Beurle, K., Brahic, A., Burns, J. A., . . . West, R. (2005). Imaging of Titan from the Cassini spacecraft. *Nature*, 434(7030), 159–168. doi: 10.1038/Nature03436.

54. Schneider, T., Graves, S. D. B., Schaller, E. L., & Brown, M. E. (2012). Polar Methane Accumulation and Rainstorms on Titan from Simulations of the Methane Cycle. *Nature*, 481(7379), 58–61. doi: 10.1038/Nature10666.

55. Tomasko, M. G., Archinal, B., Becker, T., Bezard, B., Bushroe, M., Combes, M., . . . West, R. (2005). Rain, Winds and Haze during the Huygens Probe's Descent to Titan's Surface. *Nature*, 438(7069), 765–778. doi: 10.1038/Nature04126.

56. Bird, M. K., Allison, M., Asmar, S. W., Atkinson, D. H., Avruch, I. M., Dutta-Roy, M., . . . Tyler, G. L. (2005). The Vertical Profile of Winds on Titan. *Nature, 438*, 800–892. doi: 10.1038/nature04060.

57. Flasar, F. M., Achterberg, R. K., Conrath, B. J., Gierasch, P. J., Kunde, V. G., Nixon, C. A., . . . Wishnow, E. H. (2005).

Titan's Atmospheric Temperatures, Winds, and Composition. *Science*, 308(5724), 975–978. doi: 10.1126/Science.1111150.

58. Kostiuk, Kostiuk, T., Livengood, T. A., Sonnabend, G., Fast, K. E., Hewagama, T., Murakawa, K., . . . Witasse, O. (2006). Stratospheric Global Winds on Titan at the Time of Huygens Descent. *Journal of Geophysical Research-Planets*, 111(E7). Artn E07s03 doi: 10.1029/2005je002630.

59. McKay, C. P., Pollack, J. B., & Courtin, R. (1991). The Greenhouse and Anti-Greenhouse Effects on Titan. *Science* 253, 1118–1121.

60. Aharonson, O., Hayes, A. G., Lunine, J. I., Lorenz, R. D., Allison, M. D., & Elachi, C. (2009). An Asymmetric Distribution of Lakes on Titan as a Possible Consequence of Orbital Forcing. *Nature Geoscience*, 2(12), 851–854. doi: 10.1038/Ngeo698.

61. Cruikshank, D. P., ed. (1995). *Neptune and Triton*. University of Arizona Press, Tucson, 1249 pp.

62. Hansen, C. J., & Paige, D. A. (1996). Seasonal Nitrogen Cycles on Pluto. *Icarus*, 120(2), 247–265.

63. Ingersoll, A. P. (1990). Dynamics of Triton's Atmosphere. *Nature*, 344(6264), 315–317.

64. Kirk, R. L., Soderblom, L. A., Brown, R. H., Kieffer, S. W., & Kargel, J. S. (1995). Triton's Plumes: Discovery, Characteristics, and Models. In *Neptune and Triton*, ed. D. P. Cruikshank. University of Arizona Press, Tucson, pp. 949–989.

65. McGrath, M. A., Lellouch, E., Strobel, D. F., Feldman, P. D., & Johnson, R. E. (2004). Satellite Atmospheres. In *Jupiter: The Planet, Satellites, and Magnetosphere*, eds. F. Baganal, T. E. Dowling, and W. B. McKinnon. Cambridge University Press, Cambridge, pp. 457–483.

66. Ingersoll, A. P. (1989). Io Meteorology—How Atmospheric Pressure Is Controlled Locally by Volcanos and Surface Frosts. *Icarus*, 81(2), 298–313.

67. Chabot, N. L., Ernst, C. M., Denevi, B. W., Harmon, J. K., Murchie, S. L., Blewett, D. T., . . . Zhong, E. D. (2012).

Areas of Permanent Shadow in Mercury's South Polar Region Ascertained by MESSENGER Orbital Imaging. *Geophysical Research Letters*, 39. Artn L09204 doi: 10.1029/2012gl051526.

68. Paige, D. A., Siegler, M. A., Zhang, J. A., Hayne, P. O., Foote, E. J., Bennett, K. A., . . . Lucey, P. G. (2010). Diviner Lunar Radiometer Observations of Cold Traps in the Moon's South Polar Region. *Science*, 330(6003), 479–482. doi: 10.1126/Science.1187726.

69. Asplund, M., Grevesse, N., Sauval, A. J., & Scott, P. (2009). The Chemical Composition of the Sun. *Annual Review of Astronomy and Astrophysics*, 47, 481–522. doi: 10.1146/Annurev.Astro.46.060407.145222.

70. Atreya, S. K. (2010). Atmospheric Moons Galileo Would Have Loved. In *Galileo's Medicean Moons: Their Impact on 400 Years of Discovery*, eds. C. Barbieri, S. Chakrabarti, M. Coradini, and M. Lazzarin. Cambridge University Press, Cambridge, pp. 130–140.

71. Guillot, T. (2005). The Interiors of Giant Planets: Models and Outstanding Questions. *Annual Review of Earth and Planetary Sciences*, 33, 493–530. doi: 10.1146/Annurev.Earth .32.101802.120325.

72. Niemann, H. B., Atreya, S. K., Carignan, G. R., Donahue, T. M., Haberman, J. A., Harpold, D. N., . . . Way, S. H. (1998). The Composition of the Jovian Atmosphere as Determined by the Galileo Probe Mass Spectrometer. *Journal of Geophysical Research-Planets*, 103(E10), 22831–22845.

73. Folkner, W. M., Woo, R., & Nandi, S. (1998). Ammonia Abundance in Jupiter's Atmosphere Derived From the Attenuation of the Galileo Probe's Radio Signal. *Journal of Geophysical Research-Planets*, 103(E10), 22847–22855.

74. Mahaffy, P. R., Niemann, H. B., Alpert, A., Atreya, S. K., Demick, J., Donahue, T. M., . . . Owen, T. C. (2000). Noble Gas Abundance and Isotope Ratios in the Atmosphere of Jupiter

from the Galileo Probe Mass Spectrometer. *Journal of Geophysical Research-Planets*, 105(E6), 15061–15071.

75. Currie, T., & Sicilia-Aguilar, A. (2011). The Transitional Protoplanetary Disk Frequency as a Function of Age: Disk Evolution in the Coronet Cluster, Taurus, and Other 1–8 Myr Old Regions. *Astrophysical Journal*, 732(1). Artn 24 doi: 10.1088/0004-637x/732/1/24.

76. Atreya, S. K., Mahaffy, P. R., Niemann, H. B., & Owen, T. C. (2002). Composition and Cloud Structure of Jupiter's Deep Atmosphere. *Highlights of Astronomy*, 12, 597–601.

77. Stevenson, D. J., & Salpeter, E. E. (1977). Phase Diagram and Transport Properties for Hydrogen-Helium Fluid Planets. *Astrophysical Journal Supplement Series*, 35(2), 221–237.

78. Ortiz, J. L., Orton, G. S., Friedson, A. J., Stewart, S. T., Fisher, B. M., & Spencer, J. R. (1998). Evolution and Persistence of 5-Micron Hot Spots at the Galileo Probe Entry Latitude. *Journal of Geophysical Research-Planets*, 103(E10), 23051–23069.

79. Atreya, S. K., & Wong, A. S. (2005). Coupled Clouds and Chemistry of the Giant Planets—A Case for Multiprobes. *Space Science Reviews*, 116(1–2), 121–136. doi: 10.1007/S11214-005-1951-5.

80. Ingersoll, A. P. (1999). Atmospheres of the Giant Planets. In *The New Solar System, Fourth Edition*, ed. J. K. Beatty, C. C. Peterson, and A. Chaikin. Sky Publishing Company, Cambridge, Massachusetts, and Cambridge University Press, Cambridge, UK, pp. 201–220.

81. Ingersoll, A. P. (1990). Atmospheric Dynamics of the Outer Planets. *Science*, 248(4953), 308–315.

82. Ingersoll, A. P., & Porco, C. C. (1978). Solar Heating and Internal Heat Flow on Jupiter. *Icarus*, 35(1), 27–43.

83. Little, B., Anger, C. D., Ingersoll, A. P., Vasavada, A. R., Senske, D. A., Breneman, H. H., . . . Team, G. S. (1999). Galileo Images of Lightning on Jupiter. *Icarus*, 142(2), 306–323.

84. Dowling, T. E., & Ingersoll, A. P. (1988). Potential Vorticity and Layer Thickness Variations in the Flow around Jupiters Great Red Spot and White Oval BC. *Journal of the Atmospheric Sciences*, 45(8), 1380–1396.

85. Ingersoll, A. P., & Cuong, P. G. (1981). Numerical Model of Long-Lived Jovian Vortices. *Journal of the Atmospheric Sciences*, 38(10), 2067–2076.

86. Marcus, P. S. (1988). Numerical-Simulation of Jupiter's Great Red Spot. *Nature*, 331(6158), 693–696.

87. Williams, G. P., & Wilson, R. J. (1988). The Stability and Genesis of Rossby Vortices. *Journal of the Atmospheric Sciences*, 45(2), 207–241.

88. Dowling, T. E., & Ingersoll, A. P. (1989). Jupiter's Great Red Spot as a Shallow-Water System. *Journal of the Atmospheric Sciences*, 46(21), 3256–3278.

89. Polvani, L. M., Wisdom, J., Dejong, E., & Ingersoll, A. P. (1990). Simple Dynamic Models of Neptune Great Dark Spot. *Science*, 249(4975), 1393–1398.

90. Vasavada, A. R., & Showman, A. P. (2005). Jovian Atmospheric Dynamics: An Update after Galileo and Cassini. *Reports on Progress in Physics*, 68(8), 1935–1996. doi: 10.1088/0034-4885/68/8/R06.

91. Rhines, P. B. (1975). Waves and Turbulence on a Beta-Plane. *Journal of Fluid Mechanics*, 69(Jun10), 417–443.

92. Williams, G. P. (1979). Planetary Circulations 2. Jovian Quasi-Geostrophic Regime. *Journal of the Atmospheric Sciences*, 36(5), 932–968.

93. Salyk, C., Ewald, S., Ingersoll, A. P., Lorre, J., & Vasavada, A. R. (2006). The Relationship between Eddies and Zonal Flow on Jupiter. *Icarus*, 185, 430–442.

94. Li, L., Ingersoll, A. P., Vasavada, A. R., Porco, C. C., Del Genio, A. D., & Ewald, S. P. (2004). Life Cycles of Spots on Jupiter from Cassini Images. *Icarus*, 172(1), 9–23. doi: 10.1016/J.Icarus.2003.10.015.

95. Kaspi, Y., Flierl, G. R., & Showman, A. P. (2009). The Deep Wind Structure of the Giant Planets: Results from an Anelastic General Circulation Model. *Icarus*, 202(2), 525–542. doi: 10.1016/J.Icarus.2009.03.026

96. Atkinson, D. H., Ingersoll, A. P., & Seiff, A. (1997). Deep Winds on Jupiter as Measured by the Galileo Probe. *Nature*, 388(6643), 649–650.

97. Schneider, T., & Liu, J. J. (2009). Formation of Jets and Equatorial Superrotation on Jupiter. *Journal of the Atmospheric Sciences*, 66(3), 579–601. doi: 10.1175/2008jas2798.1.

98. Gierasch, P. J., Conrath, B. J., & Magalhaes, J. A. (1986). Zonal Mean Properties of Jupiter Upper Troposphere from Voyager Infrared Observations. *Icarus*, 67(3), 456–483.

99. Porco, C. C., West, R. A., McEwen, A., Del Genio, A. D., Ingersoll, A. P., Thomas, P., . . . Vasavada, A. R. (2003). Cassini Imaging of Jupiter's Atmosphere, satellites, and Rings. *Science*, 299(5612), 1541–1547.

100. Ingersoll, A. P., Gierasch, P. J., Banfield, D., Vasavada, A. R., & Team, G. I. (2000). Moist Convection as an Energy Source for the Large-Scale Motions in Jupiter's Atmosphere. *Nature*, 403(6770), 630–632.

101. Sanchez-Lavega, A., & Battaner, E. (1987). The Nature of Saturn's Atmospheric Great White Spots. *Astronomy and Astrophysics*, 185(1–2), 315–326.

102. Fletcher, L. N., Hesman, B. E., Irwin, P. G. J., Baines, K. H., Momary, T. W., Sanchez-Lavega, A., . . . Sotin, C. (2011). Thermal Structure and Dynamics of Saturn's Northern Springtime Disturbance. *Science*, 332(6036), 1413–1417. doi: 10.1126/Science.1204774.

103. Fischer, G., Kurth, W. S., Gurnett, D. A., Zarka, P., Dyudina, U. A., Ingersoll, A. P., . . . Delcroix, M. (2011). A Giant Thunderstorm on Saturn. *Nature*, 475(7354), 75–77. doi: 10.1038/Nature10205.

104. Sanchez-Lavega, A., del Rio-Gaztelurrutia, T., Hueso, R., Gomez-Forrellad, J. M., Sanz-Requena, J. F., Legarreta, J., . . . Team, I. O. P. W. (2011). Deep Winds beneath Saturn's Upper Clouds from a Seasonal Long-Lived Planetary-Scale Storm. *Nature*, 475(7354), 71–74. doi: 10.1038/Nature10203.

105. Sayanagi, K. M., Dyudnia, U. A., Ewald, S. P., Fischer, G., Ingersoll, A. P., Kurth, W. S., Muro, G. D., Porco, C. C., West, R. A. (2012). Dynamics of Saturn's Great Storm of 2010–2011 from Cassini ISS and RPWS. *Icarus*, 223, 460–478. doi: 10.1016/J.Icarus2012.12.013.

106. Dyudina, U. A., Ingersoll, A. P., Ewald, S. P., Porco, C. C., Fischer, G., Kurth, W. S., & West, R. A. (2010). Detection of Visible Lightning on Saturn. *Geophysical Research Letters*, 37. Artn L09205 doi: 10.1029/2010gl043188.

107. Smith, M. D., & Gierasch, P. J. (1995). Convection in the Outer Planet Atmospheres Including Ortho-Para Hydrogen Conversion. *Icarus* 116, 159–179.

108. Gurnett, D. A., Groene, J. B., Persoon, A. M., Menietti, J. D., Ye, S. Y., Kurth, W. S., . . . Lecacheux, A. (2010). The Reversal of the Rotational Modulation Rates of the North and South Components of Saturn Kilometric Radiation near Equinox. *Geophysical Research Letters*, 37. Artn L24101 doi: 10.1029/2010gl045796.

109. Smith, B. A., Soderblom, L., Batson, R., Bridges, P., Inge, J., Masursky, H., . . . Suomi, V. E. (1982). A New Look at the Saturn System—The Voyager-2 Images. *Science*, 215(4532), 504–537.

110. Anderson, J. D., & Schubert, G. (2007). Saturn's Gravitational Field, Internal Rotation, and Interior Structure. *Science*, 317(5843), 1384–1387. doi: 10.1126/Science.1144835.

111. Read, P. L., Dowling, T. E., & Schubert, G. (2009). Saturn's Rotation Period from Its Atmospheric Planetary-Wave Configuration. *Nature*, 460(7255), 608–610. doi: 10.1038/Nature08194.

112. Sromovsky, L. A., Fry, P. M., Hammel, H. B., Ahue, W. M., de Pater, I., Rages, K. A., . . . van Dam, M. A. (2009). Uranus at Equinox: Cloud Morphology and Dynamics. *Icarus*, 203(1), 265–286. doi: 10.1016/J.Icarus.2009.04.015.

113. Sromovsky, L. A., Fry, P. M., Dowling, T. E., Baines, K. H., & Limaye, S. S. (2001). Neptune's Atmospheric Circulation and Cloud Morphology: Changes Revealed by 1998 HST Imaging. *Icarus*, 150(2), 244–260. doi: 10.1006/Icar.2000.6574.

114. Kliore, A., Levy, G. S., Cain, D. L., Eshleman, V. R., Fjeldbo, G., & Drake, F. D. (1965). Mariner 4 Measurements near Mars—Initial Results—Occultation Experiment—Results of 1st Direct Measurement of Mars Atmosphere and Ionosphere. *Science*, 149(3689), 1243–1248.

115. Hide, R., & Ibbetson, A. (1966). An Experimental Study of Taylor Columns. *Icarus*, 5(3), 279–290.

116. Peale, S. J., Cassen, P., & Reynolds, R. T. (1979). Melting of Io by Tidal Dissipation. *Science*, 203(4383), 892–894.

117. Kivelson, M. G., Khurana, K. K., Russell, C. T., Volwerk, M., Walker, R. J., & Zimmer, C. (2000). Galileo Magnetometer Measurements: A Stronger Case for a Subsurface Ocean at Europa. *Science*, 289(5483), 1340–1343.

Further Reading

GENERAL INTRODUCTIONS TO PLANETS AND CLIMATE

Beatty, J. K., Peterson, C. C., & Chaikin, A., eds. (1999). *The New Solar System, Fourth Edition*. Cambridge University Press, 421 pp.

Kasting, J. (2010). *How to Find a Habitable Planet*. Princeton University Press, Princeton, 326 pp.

Randall, D., (2012). *Atmosphere, Clouds, and Climate*. Princeton University Press, Princeton, 277 pp.

TEXTBOOK ON PLANETARY SCIENCE

de Pater, I., & Lissauer, J. J. (2010). *Planetary Sciences, Second Edition*. Cambridge University Press, Cambridge, 647 pp.

TEXTBOOKS ON ATMOSPHERIC SCIENCE

Salby, M. L. (1996). *Fundamentals of Atmospheric Physics*. Academic Press, San Diego, 627 pp.

Holton, J. R. (2004). *An Introduction to Dynamic Meteorology, Fourth Edition*. Elsevier Academic Press, Amsterdam, 535 pp.

COMPREHENSIVE TEXTBOOK ON ATMOSPHERIC AND OCEANIC FLUID DYNAMICS

Vallis, G. K. (2006). *Atmospheric and Oceanic Fluid Dynamics.* Cambridge University Press, Cambridge, 745 pp.

COMPREHENSIVE REFERENCE BOOKS ABOUT THE PLANETS

Baganal, F., Dowling, T. E., & McKinnon, W. B., eds. (2004). *Jupiter: The Planet, Satellites, and Magnetosphere.* Cambridge University Press, Cambridge, 719 pp.

Bougher, S. W., Hunten, D. M., & Phillips, R. J., eds. (1997). *Venus II.* University of Arizona Press, Tucson, 1362 pp.

Brown, R., Lebreton, J.-P., & Waite, J. H., eds. (2009). *Titan from Cassini-Huygens.* Springer Dordrecht, Heidelberg, 535 pp.

Cruikshank, D. P., ed. (1995). *Neptune and Triton.* University of Arizona Press, Tucson, 1249 pp.

Dougherty, M. K., Esposito, L. W., & Krimigis, S. M., eds. (2009). *Saturn from Cassini-Huygens.* Springer Dordrecht, Heidelberg, 805 pp.

Kieffer, H. H., Jakosky, B. M., Snyder, C. W., & Matthews, M. S., eds. (1992). *Mars.* University of Arizona Press, Tucson, 1498 pp.